T0251368

THE BALANCED ENGINEER

IEEE ◆ USA

THE BALANCED ENGINEER

Essential Ideas for Career Development

1998 Professional Activities Conference Proceedings
Phoenix, Arizona ▼ September 4-7, 1998

CRC Press
Taylor & Francis Group
Boca Raton London New York

CRC Press is an imprint of the
Taylor & Francis Group, an **informa** business

Technomic Publishing Company, Inc.
851 New Holland Avenue, Box 3535
Lancaster, Pennsylvania 17604

Main entry under title:
   The Balanced Engineer: Essential Ideas for Career Development

Library of Congress Catalog Card No. 98-86723
ISBN No. 0-87942-324-2

Table of Contents

Foreword   XV

HOW TO CONDUCT EFFECTIVE PERFORMANCE APPRAISALS: COMPETENCY-BASED PERFORMANCE MANAGEMENT 1
O. BLAKE

Competency-Based Performance Management   2
Some Do's and Don'ts of Performance Management   4
Accountability—The Reason Behind It All   6
Accountability Is Misunderstood   6
Why Performance Management?   7
Performance Management and Appraisal Process   8
Conceptualizing in Terms of Results   8
Objectives of Performance Management and Appraisal   8
About the Author   9

EVERYTHING YOU EVER WANTED TO KNOW ABOUT STATE LEGISLATURES, BUT WERE AFRAID TO ASK 10
C. BRANTLEY

About the Author   18

GRASS-ROOTS EFFORTS THAT CAN AFFECT PUBLIC POLICY . . . 19
Effective Lobbying at the Grass-Roots Level   19
C. BRANTLEY

Abstract   19
Why Grass-Roots?   19
Why Me?   20
How to Build a Local Grass-Roots Network   21
The Tools of the Trade   22
What IEEE-USA Is Doing About Grass Roots   26
Conclusion   29
About the Author   29

Recommendations for Effective Communications to Influence Federal Policy Issues of Importance to Electrical Engineers   30
R. M. PAUL

Abstract   30
Why Should I Care?   30
You Have Decided to Become Involved in the Federal Policy Process.
   Now What?   31
Learning to Talk the Talk   31
Content Preparation and Organizing Your Thoughts: Build a Story   31
Conclusion   33
About the Author   33

EMPLOYERS' ENGINEERING EDUCATION NEEDS FOR THE NEW MILLENNIUM **34**
L. E. BRYANT

Abstract   34
Background   34
Vision for New Engineers   35
About the Author   35

DEVELOPING AND MAINTAINING A COMPETITIVE CAREER **36**
K. BUCKNER

Abstract   36
Introduction   37
Career Bests: Where Development Occurs   37
Identifying Individual Interests   39
Identifying the Organization's Needs   41
Checking for Reality   43
Individual Development in Action: A Case Study   44
Conclusion   45
About the Author   46
About the Presenter   46

EFFECTIVE COMMUNICATION SKILLS FOR ENGINEERS **47**
S. CERRI

Abstract   47
Introduction   48
What Is Communication Excellence?   48
Humans as Satellites   49
Who Is Responsible for Effective Communication?   49

How Do Engineers and Managers Communicate Differently?   49
Communication Pitfalls: Deletion, Distortion, and Generalization   50
Communication "Representational Systems"   50
The Three Ways We Communicate 100 Percent of the Time   50
Building Rapport—How Friends Communicate   51
The "7-Step Communication Process"   51
Summary   52
References   52
About the Author   53

**TRANSITIONING FROM TECHNICAL PROFESSIONAL
TO MANAGER** **54**
S. CERRI

Abstract   54
What Is a Technical Professional?   55
What Is Important?   55
What Does It Take to Motivate People?   55
A Question of Sorting   56
Summary   57
References   57
About the Author   57

**SUCCESSFULLY SPEAKING: WINNING GOVERNMENT ORALS BY
GIVING MEMORABLE SPEECHES** **59**
R. L. CRANSTON

Introduction   59
Engineers Must Build Public Speaking Skills to Survive   59
Planning and Practice Are Key   62
Conclusion   63

**COACHING FOR CONTRIBUTION: THE LEADERSHIP BEHAVIORS
THAT MAKE A DIFFERENCE** **65**
R. CUTADEAN

Abstract   65
Introduction   65
Coaching Defined   66
Collaboration Builds "Win-Win" Opportunities   66
A Framework for Coaching   67
Establish a Climate for Learning and Trust   68
Create a Context for Performance and Development   69
Promote Discovery and Self-Reliance   70

Delegate and Instruct   71
Communicate Expectations and Mirror Reality   71
Extend Self-Reliance and Build Networks   72
Conclusion   73
About the Author   73

MIXED-SEX TEAM COMMUNICATION: AVOIDING NEGATIVE CONFLICT 74
A. ECKSTAT

Abstract   74
Introduction   74
Male-Female Communication Differences   76
Conclusion   78
Summary   79
References   79
About the Author   80

DEALING WITH OFFICE POLITICS 81
W. C. EGGERS

Abstract   81
Politics—What Is It?   81
Politics—So What?   82
Politics—Can We Do Anything?   82
Effective Communications—Good Politics?   82
Planning—Good Politics?   85
Summary   87
Bibliography   87
About the Author   88

IEEE ENGINEERING MANAGEMENT SOCIETY: A RESOURCE FOR YOUR IEEE RESPONSIBILITIES AND YOUR CAREER 89
G. GAYNOR and C. VOEGTLI

Abstract   89
EMS Membership Benefits   90
Publications   90
Education   91
Awareness   93
Networking   93
Chapter Operations   94
The Future of EMS   94
The Relevance of EMS   95
About the Authors   95

BASICS OF INTELLECTUAL PROPERTY: PATENTS, COPYRIGHTS AND TRADEMARKS 97
A. H. GESS

Abstract   97
Introduction   97
Patents   98
The Patent Application   98
Examination Process   100
Cashing In   100
Copyright   101
Requirements of Copyright   101
Protection Provided   102
Registration   102
Authorship and Ownership   103
Everyday Acts of Infringement   104
Trademarks   105
Trademark Origins   105
Protection for the Public   106
Protection for the Businesses   106
Picking a Trademark   106
Federal Registration   107
About the Author   108

COMMUNICATING WITH YOUR CUSTOMERS: GETTING INTO YOUR CUSTOMERS' WORLDS TO SERVE THEM BETTER 109
J. HARPER

Abstract   109
Introduction   110
Communicating with Your Customer   111
Active Listening   111
Empathy   112
Obtaining Customer Input   113
Summary   113
About the Author   114

LIFE WORK: REDUCING STRESS AND PLANNING YOUR CAREER TOWARD PERSONAL VALUES 115
J. A. HOSCHETTE

Abstract   115
Introduction   115
Developing a Career Mission Statement, Goals, and a Strategic Plan   116

Self-Evaluation and Identification of Values   116
Mission Statement and Goals   118
Action!   123
About the Author   124

INNOVATE FOR TODAY 125
B. KRAUSE

Abstract   125
Innovation   125
Principle of Innovation   126
Recent Innovations   126
Needing Innovation   128
About the Author   129

YES, "TEKKIES" CAN TALK—AND SOMETIMES EVEN SELL 130
T. LEECH

Abstract   130
Engineers and Persuasive Communication—An Unlikely Match?   130
Ten Tips for Becoming a Persuasive Presenter   131
Summary   136
About the Author   136

PEOPLE SKILLS IN A COMPETITIVE ENVIRONMENT 137
J. V. LILLIE

Abstract   137
The Competitive Environment   137
Skills Identification   138
Communications   138
Teamwork   139
Negotiating   140
Manners   140
Ethics   140
Attitude   141
Humor   141
Skills Assessment   141
Summary   142
About the Author   142

THE YOUNG PROFESSIONAL AS MANAGER: MANAGING OLDER SUBORDINATES 144
S. LOCKHEAD

Abstract   144
Introduction   144

Establishing Yourself   145
Establishing Service Levels   146
Empowerment   147
Out of the Cubicle Thinking   147
Action Items   148
Conclusion   149
References   149
About the Author   150

TRENDS IN EMPLOYEE BENEFITS **151**
G. F. McCLURE

Abstract   151
Introduction   151
The Menu   151
Where Benefit Dollars Are Spent   153
Current Benefits Issues   153
Flex Plans   154
Pension Issue   156
Bibliography   156
About the Author   157

IS TAX RELIEF REAL? **158**
G. F. McCLURE

Abstract   158
Introduction   158
The Breaks   158
Education   159
Retirement Savings Incentives   160
Withdrawal Penalty Repealed   160
Roth IRA   160
Capital Gains Tax Break   161
Sale of Residence   161
Estate Taxes   161
Home Office Deduction   162
Health Insurance Premiums   162
Alternative Minimum Tax   162
Underpayment Penalties   163
Bibliography   164
About the Author   165

NETWORKING: MULTI-LEVEL MARKETING YOURSELF **166**
D. J. PIERCE

Reference   169

About the Author   169
Networking Exercise   170

**BENCHMARKING ENGINEERING SKILLS AGAINST A
RAPIDLY CHANGING FUTURE** 171
L. VAVRA

Abstract   171
Introduction   171
Betting on the Future   172
The Program: A Three-Step Process   172
Summary   173
Workshop References   173
About the Author   174

**CAREER MANAGEMENT AS PERSONAL MARKETING AND
BUSINESS DEVELOPMENT** 175
C. VOEGTLI

Abstract   175
Introduction: Understanding Your Customer   175
The Skills We Need and Why   180
Developing Your Skill Set   183
How to Market Yourself Continually and Develop Future "Business"   187
Conclusion   189
About the Author   189

**LESSONS IN CAREER MANAGEMENT FROM SILICON VALLEY:
KEY FACTORS FOR CONTINUOUS "EMPLOYMENT"** 191
C. VOEGTLI

Abstract   191
The Current Outlook for Consultants   192
Successes and Failures in the Consulting World   193
The Important Capabilities   196
How to Develop These Skills   199
Conclusion   199
About the Author   200

**EFFECTS OF HUMOR ON SOCIAL INFLUENCE STRATEGIES IN
A WORKPLACE SCENARIO** 201
M. WALTERS

Abstract   201
Introduction   201

Humor   201
Power Bases   203
Humor and Influence   204
Hypotheses   204
Method   205
Results   207
Discussion   209
Conclusion   211
References   212
About the Author   212

Author Index   215

Foreword

This volume constitutes the Proceedings of the 1998 IEEE-USA Professional Activities Conference, the second annual professional activities conference sponsored by The Institute of Electrical and Electronics Engineers-United States of America. These national meetings are a continuation of the PACE Conferences held annually since 1976 by the Professional Activities Committees for Engineers (PACE) of the former IEEE United States Activities Board.

The theme of the 1998 Professional Activities Conference is "Preparing for the New Millennium." The event is open to all professionals. Our goal is to assist individuals with the development of leadership, teamwork, negotiating, networking, and other professional skills. IEEE members who attend should develop into more effective PACE leaders and IEEE volunteers and more productive employees.

I welcome you to the conference and encourage you to participate fully. Attend as many sessions as possible. Take advantage of the networking opportunities offered at the sessions and the social activities.

As chair of the 1998 conference, I would like to thank everyone who assisted in planning and presenting the event. No conference can be successful without the collaboration of a dedicated team. My congratulations for a job well done go to all the members of the Operations Committee and its subcommittees. Charles Lessard, chair of several past conferences, provided valuable assistance as my conference co-chair. George McClure, chair of the Program Subcommittee, coordinated all aspects of establishing a great program. Hollis Hart and his GOLD (Graduates of the Last Decade) Subcommittee, ensured the participation of a large number of young professionals and organized two tracks of sessions dedicated to their interests. Glenda McClure did a fantastic job of developing and hosting the Companion Program. Through his leadership, IEEE-USA President John Reinert ensured, as always, that all things would work well. Committee members Mike Andrews, Jean Eason, Robert Gauger, Leann Kostek, Loren Lacy, Robert Powers, Lee Stogner, and Ed Wong also made significant contributions. Administrative support from IEEE-USA staff and IEEE Meeting Planning Services provided the glue that held the operation together.

Please utilize these proceedings as a reference to improve your nontechnical skills and increase your value to your family and your employer. In addition, please share this document with others who have a need for skill improvement.

Enjoy the conference! And make plans now to attend the 1999 IEEE-USA Professional Development Conference, which will be held in Dallas, September 3–6, 1999.

JOSEPH V. LILLIE
Chair, 1998 Professional Activities Conference and
IEEE-USA Vice President—Professional Activities

How to Conduct Effective Performance Appraisals: Competency-Based Performance Management

O. BLAKE

Definition: *Performance appraisal is the process of defining critical, position requirements by which expectations, standards, evaluations, documentation, and feedback may be communicated to each person.*

Loved and hated equally, performance management programs (PMPs) have always sparked debate about their effectiveness. Proponents have claimed that PMPs boost the bottom line by linking employee accountabilities to organizational goals. Critics maintain that PMPs encourage mediocrity by rewarding easily attainable goals. In the worst case, critics assert that PMPs create irreversible conflict between supervisors and employees.

A study reported in *Harvard Business Review* (September–October 1996) suggests that companies that use PMPs perform better financially than companies that don't. Danielle McDonald, Katie Donohue, and Sally Shield of Hewitt Associates and Abbie Smith at the University of Chicago conducted the research, which was based on an analysis of 437 publicly held U.S. companies from 1990 through 1994. The research indicates that PMPs encourage employees and managers to channel their time and energy toward defined strategic goals.

To clarify, sitting down with people to set goals, coach them, and give them feedback about their performance creates an opportunity for managers to help people focus on the kinds of activities that will help the company realize its objectives.

When people can see the bigger picture and understand how their contribution fits with the greater goals of the company, they will do a better job.

The researchers found that successful performance management programs shared several characteristics:

1. Implementation at the top—senior management helped design the program
2. A simple process—the program was integrated easily into employees' daily work
3. A reasonable number of goals—limited to four or five
4. Additional feedback—performance programs were supplemented by *informal regular feedbacks*

Orlando (Lanny) Blake, The Blake Group, Glendora, California

Once in place, PMPs must be continually evaluated and changed appropriately as the organization changes. However, the researchers say that it is worth the effort to keep PMPs current!

COMPETENCY-BASED PERFORMANCE MANAGEMENT

Competencies are the skills, knowledge, abilities, characteristics, and other attributes that, in the right combination, and for the right set of circumstances, tend to predict superior performance.

Performance competencies are job-related behaviors and characteristics that are requirements for the successful employee. That is, in order for an employee to be successful, he or she must be able to behave in specific ways and at acceptable levels over time. These behaviors are functional in the sense that they are transferable to many jobs within the world of work. The difference will vary by degree in relevance to your organization and job for which you are appraising performance.

The defined performance competencies are related to both successful performance and higher earnings for the people who possess them. Employers and people need to develop these skills and competencies actively, if they are going to meet the demands of a high-performance workplace.

Performance competencies are identified and weighted according to criticality of successful job performance.

Competency-based elements may be linked to any or all of the key components of the total reward package:

- Base salary
- Salary increases
- Variable pay
- Training and development
- Succession planning

How to Identify Critical Performance Competencies

Ways to discover competencies were proposed by Jim Kochanski in "Competency-Based Management," published in *Training & Development* (October 1997, pp. 41–46).

Approach	Strength	Limitation
Analyze a star performer	Reveals secrets of the pros	Job-specific; complex architecture

Approach	Strength	Limitation
Analyze many exemplary employees	Easy to generalize; simple architecture	Time-consuming
Survey experts	Fast, statistically valid	Low touch; lack of buy-in
Compile external models	Garners best of the best	Not necessarily applicable to the organization

There are six ways to identify competencies:

1. List accountabilities
   A) Responsibilities
   B) Authority
   C) Identify results of the most critical areas of performance
2. Use a questionnaire
   A) If geographically dispersed
   B) If duties are highly complex
   C) Questions to ask
      1) What is the purpose of the work?
      2) What am I trying to make happen?
      3) What conditions would exist if work were performed satisfactorily?
      4) What would the company lose if work were performed unsatisfactorily?
      5) How does the client define success?
3. Use role charting
4. Use a process-coded responsibility chart
5. Q-sort responsibility chart
6. Use the accountability statements from the job description

You can spend a lot of time creating competency models. Are they worth it? Maxine Dalton ("Are Competency Models Worth It?" *Training & Development*, October 1997, pp. 46–49) reports that all models of effective leadership can be factored into major categories: cognitive skills, interpersonal skills, personal skills, and knowledge of the business. She asserts that companies can save a lot of time and money by using an existing research-based management model to derive human resource development strategies. She suggests that organizations that lack resources to create and validate their own competency model can adapt an off-the-shelf model that is based in research.

Many such models exist. For example, this author has developed the Critical Attributes of Performance System (CAPs™), based on his research. CAPS™ is part of a comprehensive performance management and employee selection system titled Selection and Assessment Interview Training (SAIT™). The employee selection part of the system is titled Behaviorally Anchored Selection System (BASS™).

CAPS™ and BASS™ were developed from data generated by 146 employers and

219 interviews. Job analyses were conducted to determine the critical attributes of success or "know how" in a variety of positions. Definitions of each attribute were refined and used to establish baseline performance competencies. Performance competencies are made up of eight skills and personal qualities needed for solid job performance.

Effective people can use the following productively:

- Resources—they know how to allocate time, money, materials, space, and staff.
- Interpersonal Skills—they can work on teams, teach others, serve customers, lead, negotiate, and work well with other people from culturally diverse backgrounds.
- Information—they can acquire and evaluate data, organize and maintain files, interpret and communicate, and use computers to process information.
- Systems—they understand social, organizational, and technological systems. They can monitor and correct performance and they can design or improve systems.
- Technology—they can select equipment and tools, apply technology to specify tasks, and maintain and troubleshoot equipment.

Competent people in the high-performance workplace need:

- Basic Skills—reading, writing, arithmetic, mathematics, speaking, and listening.
- Thinking Skills—the ability to learn, to reason, to think creatively, to make decisions, and solve problems.
- Personal Qualities—individual responsibility, self-esteem, self-management, sociability, and integrity.

The performance competencies defined here are related to both successful performance and higher earnings for the people who possess them. Employers and people need to develop these skills and competencies actively, if they are going to meet the demands of a high-performance workplace. There are 35 performance competencies to identify and weigh according to criticality of successful job performance.

SOME DO'S AND DON'TS OF PERFORMANCE MANAGEMENT

Don'ts

1. Don't try to measure everything. Focus on the most critical aspects and understand that everyone has routine tasks.
2. Don't try to have too many performance competencies. If there are too many, the job may be too much for one person or some things may not be getting done.
3. Don't try to make the measures too precise. The measures must make sense to everyone and should not be so rigid that flexibility is lost.
4. Don't keep the methods of measurement a secret. If the measurement method is arrived at mutually, the chances of mutual commitment are high and there will be less chance of game-playing or cheating.

5. Don't measure subjective, judgmental traits, such as attitude, character, appearance, enthusiasm, cooperation, or initiative.

Do's

1. Always consider quality, quantity, time, and cost.
2. Attempt to develop the measures mutually, either as a work group or one-on-one with each person.
3. Measure results and outputs rather than methods or activity.

YES	NO
Cost per unit of hire	Number of interviews
Cost per unit of output	Number of units produced
Percentage of quality problems solved at plant	Number of samples taken

4. State ranges of the results, rather than results. For example:
   Minimum   20% sales increase
   Maximum   35% sales increase
   Achievable   28% sales increase
5. If necessary use percentages (i.e., percentage of cases won versus number filed); ratios (i.e., number of hires versus number of retentions); or indexes (i.e., attendance − accidents − overtime − turnover = .83)
6. Try for trade-offs between mutually exclusive aims or measures, such as with "I want to increase production, improve quality, reduce labor and decrease cost."
7. Make critical probes to evaluate trade-offs. For example, "What remaining aspects of performance deteriorate by focusing improvement on this one?" or, "If schedules were prepared to implement these objectives, what time utilization and critical path among other activities would be required?" or "What sequential steps are needed to accomplish this goal?" or "What could go wrong or keep you from accomplishing your goal?"
8. Where the value of performance is abstract, initiate practices that will make it measurable. For example, how can you measure overhead or staff functions (personnel, legal, public relations, research, I.E., marketing, project engineering)?
9. If you cannot predict conditions upon which performance success depends (i.e., rapid change, project work, sequential interdependence) use floating or gliding goals. Examples include: amount produced as a percentage of schedule; collections as a percentage of sales; efficiency as a percentage of learning curve; percentage of

quality control problems solved (re-inspections); percentage of project milestones; and percentage of delivery schedules met (dates).

ACCOUNTABILITY—THE REASON BEHIND IT ALL

Many people ask, "Why go to all this extra work? Why bother with all of this? The job is getting done. Is it really worth the effort?" The answers lie in an organization's value system regarding accountability. Performance Management (PM) has the potential to increase productivity, but that is relative to being accountable for an agreed upon result. Unless individuals or organizations are accountable, there is little promise or justification for seeking a result orientation. If there is no real commitment to working the accountability issue, there is little justification for engaging in Performance Management (PM). The true payoff is lost without the discipline of managing the accountability for results.

Accountability needs to be directed not only toward task or work achievement,
but also toward maintaining relationships and affirming one's self.

We are all under increased pressure for effectiveness in achieving results. This approach includes considerable emphasis on shared accountability and responsibility for contributing to and achieving results. In itself this is somewhat of a step forward compared to past management practice. But this change is part of the necessary movement forward.

Rewards based on real performance provide the solid basis for being accountable. Rather than influence or personality, accountability based on contribution to organization results leads to managing for the effective utilization of all resources available.

ACCOUNTABILITY IS MISUNDERSTOOD

People are accountable for success or failure in accomplishing objectives. Job descriptions should include a section on accountabilities that describes end results achieved when job duties are performed satisfactorily and standards for measuring performance.

Managers and supervisors should realize that when authority is delegated,
the person to whom it has been delegated is held accountable.

The person must consciously accept the responsibility to exercise authority when making and carrying out decisions.

Accountability Statements—Vice President of Human Resources: The vice president of human resources is responsible for recruiting, placement, training, compensation, benefits, safety, and security of all company employees. This office will also be responsible for compliance with all EEOC, AAP, diversity initiatives, and all other state and federal laws.

Examples:

Statement	Measure
Formulation of personnel policies that improve job satisfaction, productivity, and profit.	Eliminate need for any third party claims Turnover rate Absentee rate AAP
Formulation of human resource objective that support company's overall objectives	Employment and placement plan Personnel training and development Compensation plan

Objectives	
No third party attempts to organize	Distinguished—no attempts Commendable—one Qualified—2–3 Fair—4 Marginal—more than 4
No third party successes in organizing	Distinguished—0 Unacceptable—1

WHY PERFORMANCE MANAGEMENT?

Performance management offers companies and their employees a number of benefits. It:

1. Enables employees to understand their current responsibilities, how management views their performance, and how they might increase their effectiveness in their current positions;
2. Provides each person the opportunity to express long-range career goals and interests;
3. Provides guidance to employees on how to increase their chances for fulfilling career interests or goals;
4. Ensures that the data describing each employee's performance, potential, interests, and readiness for other positions be examined and updated at least annually and be used in making internal searches to fill open positions;
5. Guides the development of internal training opportunities;
6. Is flexible enough to accommodate the variety of functions and diversity in which it will be used;

7. Reflects the changing attitudes of both the courts and regulatory agencies in this area; and

8. Supports the need of the compensation programs for individual performance.

PERFORMANCE MANAGEMENT AND APPRAISAL PROCESS

The process of managing for performance must include:

1. Job-related performance requirements (i.e., critical performance areas and acceptable performance levels)
2. Communication (i.e., what to do, how well to do it, and how it will be measured)
3. Evaluation (i.e, actual performance versus acceptable performance)
4. Documentation (i.e., journal or log performance data and use of a common form)
5. Action (include performance feedback, a development plan, and compensation decisions)

CONCEPTUALIZING IN TERMS OF RESULTS

Start where you want to end. The ability to think and conceptualize in terms of results is essential for making PM work. To conceptualize in terms of results, people and organizations must think beyond their own boundaries. It means we are free enough to be aware of the demands that others are making of our work group or of us.

Rather than looking within to improve our own efficiency, we must first look outside to validate that our results are meeting the needs and values of the total environment in which we are functioning. This approach runs counter to most of our historical experience, which is based on a rational, mechanistic philosophy that says you should first seek to get your own work perfect and that should give you the "best" performance. But if you don't know what the "market" needs, what good is it to become super-efficient at doing something that might not be needed or is of marginal value in terms of organization performance and achievement.

OBJECTIVES OF PERFORMANCE MANAGEMENT AND APPRAISAL

An effective Performance Management Process (PMP) should be the primary support for:

1. Talent assessment, to promote the best qualified personnel.
2. Productivity assurance, to identify and provide the training necessary to ensure proficient performance, and to reward high levels of performance.

3. Talent development, to provide opportunities and job assignments to ensure full employee utilization and to identify or develop back-up strength for succession planning.

4. Protection from regulatory interference, to avoid or defend discrimination charges and to provide necessary validation data.

ABOUT THE AUTHOR

Orlando (Lanny) E. Blake, Jr., is president of The Blake Group (Glendora, California), which specializes in mediation and teamwork, leadership, and communication training and consulting, executive assessment, and coaching. Mr. Blake's career in human resources management and organization effectiveness has spanned more than 20 years and has included executive positions as senior vice president at Mercantile National Bank and as director of personnel at Warner Bros., Inc. His diverse industry experience includes printing, entertainment, pharmaceuticals, banking, food processing, apparel manufacturing and retailing, heavy engineering construction, publishing, and auto services. Before starting The Blake Group in 1993, Mr. Blake was director of human resources at GUESS? Inc., a leading multinational apparel manufacturer and retailer.

Mr. Blake can speak authoritatively about change and conflict. He has managed large-scale organizational restructuring, effectiveness programs, and work design change efforts for a variety of industries.

Mr. Blake received his master's degree in public administration from the University of Southern California, with an emphasis in applied behavioral science. His doctoral research at Claremont Graduate School emphasized organizational change and conflict resolution. His professional affiliations include the Society for Human Resources Management, the Society for Professionals in Dispute Resolution, the Southern California Mediation Association, the American Society for Training and Development, and the Association for Psychological Type.

Mr. Blake is a contributing editor with Personnel Policy Service's flagship publication, *The Personnel Policy Manual.* He writes, assists and advises the publisher on the practical application of management policies and human resources management practices. Mr. Blake continues to write, speak and teach on management effectiveness, conflict resolution, organizational change, and career transition. He has developed several proprietary instruments for selection and assessment interviewing, competency-based performance management, and career re-visioning.

Mr. Blake began his employment in the family cattle business. It is rumored that he learned his interpersonal and negotiation skills as an apprentice buyer.

Everything You Ever Wanted to Know About State Legislatures, But Were Afraid to Ask

C. BRANTLEY

This paper provides answers to 10 frequently asked questions about state legislatures and why they are important. It can't hope to cover everything you might want to know about state legislatures and how they operate, and there is also no reason why I should think you would be afraid to ask a question about that topic. What it does do, however, is attempt to explain why it is important to keep an eye on and understand what goes on in your state legislature and give you some tips on how to do that. It does so by answering 10 questions I'm asked about state legislatures in my capacity as IEEE-USA's manager of government activities.

1. Why should I care about my state legislature?

There are any number of reasons why you should care about what goes on in your state legislature, the first and foremost of which is that your state legislators are writing laws that can affect you, your family, your career and profession, and your community.

You may believe that your state legislature exists to help solve the pressing problems confronting your state and community. Or you may believe that "no one's life, liberty or property is safe while the legislature is in session." Whatever your attitude toward government and the legislative process, there is no denying that the outcome will affect your interests in some way, either directly or indirectly.

In addition to self-interest, you also have a public responsibility to be concerned about what goes on at the state legislature. As a citizen, you have a responsibility to be knowledgeable about the issues and to vote for well-qualified candidates to represent you. Moreover, as engineers, you have a professional responsibility to help ensure that your state legislature understands the technical implications of the increasingly complex subjects that it legislates, ranging from Internet taxation and electricity deregulation to regulation of telecommunications and the use of technology to provide state services.

Many individuals who follow national issues closely and are familiar with the deliberations of the U.S. Congress spend little or no time following what goes on in their state's legislature. This can be a mistake. First, laws adopted by the state legislature are more likely to affect you and your community directly than laws adopted in Washing-

Chris Brantley, Manager, Government Activities and Operations, IEEE-USA

ton. The quality of your roads, schools, and public services are primarily the subjects of state laws and local regulations, not Washington decrees. Sales and property taxes are frequently state- or locally based. States regulate professional licensure and registration. Moreover, as the national government has taken steps to balance the federal budget and shrink the size of government, more and more responsibilities are being shifted to the states, including health care and social services.

For some inexplicable reason, Americans frequently dismiss any thought of active participation in the legislative process with the excuse that their involvement is not likely to make a difference anyway. This form of political apathy is inexcusable and creates a self-fulfilling prophesy. Our democratic system of government can only function effectively with a well-informed and involved citizenry. Many people greatly underestimate the amount of access they enjoy to their representatives and the weight that their voice (and vote) can have, especially when coupled with those of other like-minded citizens. The individuals and organizations who understand the power of grass-roots politics control the legislative agenda at the state and federal levels.

Finally, remember that it's your tax money that the state legislature is spending. Certainly you have an interest in making sure that your hard-earned money is used in the most effective way possible and not wasted so that the state doesn't have any reason to come back for even more.

2. What kinds of issues do state legislatures deal with that concern engineers specifically?

IEEE-USA's State Government Activities Committee has identified the following issues as being addressed regularly at the state level. The first category concerns issues that affect the practice of engineering:

• Licensure and registration generally
• Continuing professional education/competency requirements
• Licensure and registration of software engineers
• Tax on professional services
• Tort reform/professional liability
• Volunteer protection from civil liability

The second category includes science, engineering and technology-related issues that arise regularly at the state level, often within the context of promoting economic development and building a high-tech economy within the state:

• Funding of state engineering schools
• State/local information infrastructure development
• Telecommunications regulation
• Energy utility restructuring
• Year 2000 problem
• Taxation of the internet
• Digital signatures
• Educational and other incentives for a high tech work force

The third category involves the state-federal relationship and especially state utilization of federal technology transfer and commercialization resources to promote economic development. Of related concern is the ability of states to comply with the technical requirements of federal regulations.

- Federal funding of Experimental Program to Stimulate Competitive Research (EPSCOR)
- State implementation of Clean Air Act mandates concerning alternative emissions inspection and maintenance programs
- State implementation of Energy Policy Act of 1992 mandates regarding utility planning, energy efficiency, building standards, and use of alternative fuels
- State-federal technology partnerships
- Coping with the problem of unfunded federal mandates
- Siting concerns about electric and magnetic fields health effects

Finally, opportunities exist for engineers to volunteer their technical expertise in their communities and states, such as:

- Precollege math and science education (alternative teaching certification)
- Representation by engineers on state boards/committees
- Total Quality/ISO 9000 implementation by state government/local business

3. How are state legislatures organized?

With the exception of Nebraska, which has a single (i.e., unicameral) legislative body, the states follow the federal model of a bicameral (two-house) legislature, typically with a larger House of Representatives, Assembly or House of Delegates and a smaller Senate. The largest state legislature is New Hampshire's, with 424 members. The smallest is Nebraska's, with 49 members. The name of the legislative body varies from state to state, and includes Legislature, General Assembly, Legislative Assembly, and General Court.

Typically, Senators are elected to longer (four-year) terms than Representatives/Delegates (although not in all states) and are expected to be the more deliberative body since they are (in theory) less subject to constant re-election pressures. The use of two legislative houses ensures checks and balances in the system, since each house must typically agree on legislation before it can be passed into law for signature by the governor.

4. For how long do state legislatures meet?

Most state legislatures meet in an annual or biennial session that begins in January or February and runs a number of calendar or legislative days fixed by state law. Sessions frequently vary in length from even to odd years and in years between and following elections. The practice of having a long session in odd-numbered years following the general election followed by a short session in even-numbered years is common. The types of legislation that can be considered in short sessions are often limited to budgetary and emergency measures.

Consistent with the idea of "citizen legislatures," most sessions are limited by state constitution or statute to 30, 60 or 90 days. Many states provide for special sessions to be held either between or concurrent with regular sessions for a variety of purposes. In addition, many states have an "interim study period" between sessions, when committees can meet and bills can be filed for consideration.

5. What are the differences between state legislatures and the U.S. Congress?

Apart from generic differences such as names and lengths of the legislative session and rules of procedure within the various bodies, there are two fundamental differences between your state legislature and the U.S. Congress.

The first fundamental difference concerns the respective roles of the state versus the federal lawmaker. The federal Representative or Senator is a professional lawmaker who works year-round on the people's business, is paid a significant salary, and has a full-time professional staff in Washington and in the state/district to provide support. Although it varies from state to state, state legislatures generally conform to the notion of citizen legislators. They typically receive little or no pay and so often must maintain other employment. They are not typically expected to serve year-round and generally have limited or no staff resources at their disposal. As a practical matter, this means that state legislators are much more dependent on outside sources of information and advice.

The second major difference concerns the subject matter that state legislatures and the U.S. Congress respectively legislate. The U.S. Constitution provides for federal preeminence on certain subjects (e.g. national security, foreign relations, currency, interstate commerce, etc.) but reserves other matters to the states. Thus states are primarily concerned with criminal and civil laws such as remedies for negligence and for matters concerning the public safety. In some cases, states and the federal government legislate different facits of the same subject. For example, both states and the U.S. Congress legislate education and transportation. Similarly, both are responsible for revenue (i.e., taxation) and budgetary matters, including funding appropriations to support the operations of government.

6. What is the best way to track important state legislation?

The answer to this question depends on whether you're monitoring a legislative session for bills that may be of interest to you or your group or whether you are trying to keep track of a specific bill as it moves through the legislative process.

In the first case, the challenge of monitoring all the legislation introduced in a specific state session and identifying those bills that may affect your specific interests is daunting. It requires time and energy to review the hundreds (or thousands) of bills that are introduced in a given session, much less follow them all the way through the legislative process to determine whether they have been amended and to decide which ones are likely to survive all the hurdles between introduction of a bill and its passage into law. For example, I recall doing a quick search of the California General Assembly's database during the 1997 session, which produced more than a 150 bills concerning some aspect of the Internet. I doubt more than one or two of those survived the process and became law.

One of the main reasons for hiring a lobbyist is to have someone who has the time and expertise to filter the bills introduced in a given legislative session quickly and accurately and determine what is relevant and what is not. Here is how lobbyists typically do that:

1. By using their knowledge of and experience with the legislative process in your state, and having a knowledge of your interests and profession, lobbyists focus on the legislators and committees who have jurisdiction over the issues (and bills) that will mostly likely concern you. If a bill does not originate with the leaders or among the members of a committee with jurisdiction over the bill, it is unlikely to see much action.

2. Lobbyists will also sign up for state bill services that provide copies (for a fee) of all bills that are introduced. The challenge then is to sort through the bills to eliminate those that do not affect their clients' interests.

3. To help focus on key bills, lobbyists will also subscribe to information services, such as legislative summaries, as well as "insider" publications and organization newsletters that keep tabs on the legislative session.

4. Lobbyists can't be everywhere at once or look at every bill that is introduced, so it is important to have a good network of information sources. Talking to key legislators is a great way to find out what bills are moving and what the major issues are. Effective lobbyists also share information within their own network of contacts.

5. Finally, there is the World Wide Web, which is rapidly becoming one of the most useful information resources for state activists because of the availability of analytical search tools that screen large amounts of data quickly. State legislatures are increasingly turning to the web as a means to distribute legislative information, and the public is making use of this tool.

You don't have to hire a lobbyist to track legislation; if you are willing to invest the time, you can do the same things a lobbyist does yourself. The task is a little bit easier if you are concerned with a specific piece of legislation. As a practical matter, the best way to track a specific piece of legislation is to identify and contact the sponsor of that bill regularly. The sponsor can tell you where the bill really stands and can identify supporters and opponents. To help citizens and organizations monitor legislation, almost all state legislatures have established phone contacts for bill status updates. Finally, states are using the World Wide Web more and more to provide information on bill status, although the nature and timeliness of the information varies widely from state to state. In some states, you can see the text of bills as they are introduced. In other states, you can also read amendments and/or later versions of the bill (e.g., "second readings") after they have gone through the hearing and amendment process. A few states provide legislative summaries and the text of proposed amendments that are still pending.

7. How can I influence state legislation?

The subject of effective lobbying can (and does) fill books. Here I can only provide a brief capsule summary. The first thing to note is that influencing your state legislator is not fundamentally different than influencing your Member of Congress. For that reason, I would refer you to the "Engineers' Guide to Influencing Public Policy" (*http://www.ieeeusa.org/usab/FORUM/GUIDE/*) and to my other presentation on "Effective Lobbying at the Grass-roots Level" for helpful hints.

The four keys to influence are:

- Making your case and delivering straight facts to your legislator;
- Having active allies;
- Mobilizing grass-roots action; and
- Forming a relationship and getting along well with your state legislator.

The traditional ways of making your case are with phone calls, letters and possibly even e-mail. Be sure to identify yourself as a constituent. Focus your message on one topic. Be clear about what you want your legislator to do. Be polite and never argumentative. Offer to provide facts and materials that support your point. Timing is also very important. Don't send your message or make your contact too early in the process, when the legislator is not concerned about the issue, and don't wait until after the bill has been marked up and voted on to express your concerns.

There are key times during the legislative cycle when you can increase the chances that your input will be heard: when a bill is introduced, when hearings are held on the bill, when the bill is "marked-up" by committee, when the bill is brought forward for a vote, and when the bill is presented to the governor for final signature.

Keep in mind two unique characteristics about state legislators. First, unlike a U.S. Congressman, a state legislator has little or no staff. Rather than working through staff, you will most likely be dealing with the decision-maker directly. Second, because sessions are so short (30 to 90 days) and legislators don't have staff, they are not always familiar with the issues on which they are asked to vote and must rely on the advice/guidance of their colleagues who served on the committees that reported those measures. As a consequence, you should not only share your views about a bill with your own state legislator, but also provide input to the sponsor of the measure and to the members of the committee(s) that were assigned jurisdiction over it.

You will have more impact on legislation if you can also get others to share their input. This is the idea of grass-roots lobbying. If the issue concerns you, it is likely to concern others in your profession, company, or community. Take the time to enlist their aid. Set up a phone tree to spread the burden of keeping people up to date. Recruit coordinators as your network expands. Hire a lobbyist if the issue (and your budget) merits that type of expenditure. Let people know what is happening on the issue on a regular basis so that they feel prepared to respond when you send them a request to take action.

Another important way to influence your state legislator is by establishing a personal relationship. This can begin by inviting your legislator to speak at a section or chapter meeting, by arranging a visit at the state capitol or in the local office, by participating regularly in town hall meetings and local forums, or even by volunteering to help with the legislator's next election campaign. Be helpful and provide credible information, and you will find that your state legislator is appreciative.

Finally, don't underestimate the indirect approach. A letter to the editor, an OpEd piece in the local paper, and/or a press release can give you media exposure, attract potential supporters, and catch the eyes of most legislators, who monitor the press closely. A media strategy should be an integral part of any significant grass-roots campaign.

8. If I get involved in state government activities, am I subject to rules and regulations as a "lobbyist"?

In our democratic system, individuals are encouraged to speak out, to express their opinions, to work for and give money to political candidates, and otherwise participate in the political process. However, when individuals or groups begin to apply resources in a systematic way to influence the outcome of legislation (i.e., "lobby"), there is also a public interest in full disclosure of that activity to discourage violations of campaign laws and ethical rules and ensure that our elected representatives are accountable for their actions. That is why all states have adopted statutes and/or regulations concerning "lobbying."

Activity that constitutes "lobbying" varies from state to state, according to differences in the applicable law. However, as a general rule, you are not considered to be a "lobbyist" unless you are compensated for influencing legislation by providing information or advocating a certain position. In some states, the nature and frequency of interactions with legislators may also determine whether or not you should register as a lobbyist. In New York, for example, you must spend at least $2,000 on covered lobbying activities before you are required to register.

For the typical citizen who writes a letter, meets with his or her state legislator, or sends a check to a candidate for public office, lobbyist registration is not something to be concerned about. If you have any doubts, however, the best thing to do is contact the appropriate legislative agency or commission responsible for regulating lobbying activity in your state. A list of contacts by state is provided in the appendix.

By the way, there is nothing wrong with being a lobbyist or advocate for a group or particular point of view; it's just a matter of registering and reporting your activities and abiding by certain ethical rules that govern your relationship with legislators, such as the giving of gifts, political contributions, and the like.

9. Where can I learn more about my state legislature and how state laws are made?

Many states publish useful citizens' guides and information resources on how their legislative system works and how you can influence it. These guides are usually available either free of charge or for a small fee. Some states also offer videos and/or provide information on the World Wide Web. The best way to find these resources is to start by browsing your state legislature's web site to see what is available. You can also contact me in the IEEE-USA office for state-specific titles and sources.

There are also a number of general published resources that you might find useful and that can be found in your local library or bookstore. Here is a short bibliography:

Council of State Governments, *The Book of the States* (Lexington, KY: 1996; comprehensive reference book)

Jack Davies, *Legislative Law and Process*, 2d ed. (St. Paul, MN: West Publishing, 1986; law text written by former state legislator)

Irving J. Gabelman, *State Government*, in Gabelman, ed., *The New Engineers Guide to Career Growth and Professional Awareness*, (New York, NY: IEEE Press, 1996).

William J. Keefe and Morris S. Ogul, *The American Legislative Process: Congress and the States*, 9th ed. (Des Moines, IA: Simon & Schuster, 9th ed., 1996) (basic college text)

William Lilly, III, et al., eds., *The Almanac of State Legislatures*, (Washington, DC: Congressional Quarterly, 1994)(comprehensive data on state legislative districts)

Alan Rosenthal, *The Third House: Lobbyists and Lobbying in the States*, (Washington, DC: Congressional Quarterly, 1992)

10. What can IEEE-USA do to help support members interested in getting involved in government activities in my state?

IEEE-USA can help you with state government activities in two ways: with "how-to" advice and information and with funding for special projects. IEEE-USA's State Government Activities Committee (SGAC) is a primary resource for advice and assistance. SGAC maintains "The Engineers' Handbook to State Government Activities," which can be found on-line at *http://www.ieeeusa.org/usab/FORUM/ HANDBOOK/*. You can find IEEE-USA position statements on state issues and links to other state government organizations and resources on-line on IEEE-USA's State Government Activities web page at *http://www.ieeeusa.org/usab/COMMITTEES/SGAC/*. Finally, I invite you to contact me directly with questions or for help. I have a number of references at my disposal, including the annually published *State Legislative Sourcebook*, which offer a wealth of useful information. I can also put you in touch with your state government activities coordinator or other individuals/organizations who are also following legislation and/or have similar concerns.

If you need funding to organize a state government activity within your IEEE section or chapter, you can apply through your region, section or division PACE chair for PACE Regional/Divisional Activities Support Funds. You will have to submit a proposal describing your proposed activity, outlining your budget, stating what you hope to accomplish, and providing a schedule. If funded, you must also agree to provide a report of outcomes and accounting of expenditures at the completion of your project. Some possible projects include organizing a state lobby day for visits with legislators in the state capitol, bringing a legislator to a local section or town hall meeting, developing background information on a particular issue or bill, or organizing a grass-roots network. Contact your PACE chair or consult the PACE Network webpage at *http://www.ieeeusa.org/usab/PACE* for more details.

ABOUT THE AUTHOR

Chris J. Brantley is manager of government activities and operations for the Institute of Electrical and Electronics Engineers-United States of America. A registered lobbyist, he is responsible for IEEE-USA's government fellowships and internships, state government activities, and implementation of a new IEEE-USA Grass-roots Network. Prior to joining IEEE in 1989, he was assistant to the executive director and acting director of government relations for the American Association of Engineering Societies. A complete resume is available on-line at *http://www.erols.com/brant/ brantley.html*. Mr. Brantley can be contacted by e-mail at c.brantley@ieee.org.

Grass-Roots Efforts That Can Affect Public Policy

The following two papers, written by IEEE-USA staff members Chris Brantley and Raymond Paul, offer related information about influencing public policy at the state and federal levels through grass-roots activities. The papers complement one another, and so they are presented together here.

Effective Lobbying at the Grass-Roots Level
C. BRANTLEY

ABSTRACT

This presentation is designed to provide a practical overview of how you can build an effective grass-roots network in your section, community, and/or state, to share information, and to mobilize action on the policy issues that concern you and the engineering profession. In addition to discussing the importance of grass-roots lobbying and providing instruction on techniques and methodology, it also highlights IEEE-USA's on-going efforts to build a grass-roots network to promote our legislative agenda.

WHY GRASS-ROOTS?

The answer to this question is quite simple. Politicians need votes in order to get elected (or re-elected). A strong grass-roots network of constituents represents a vote-generating (or alienating) engine that a good candidate or elected official can't afford to ignore.

For its December 1997 issue, *Fortune* magazine interviewed more than 2,000 Washington insiders, including members of Congress, their staffs, and White House officials, and asked them to rank the most influential lobbies in Washington and identify what made them influential. The answers shouldn't be surprising. Dominating the top 10 lobbies were national organizations such as the American Association of Retired Persons, the National Federation of Independent Businesses, the National Rifle Asso-

---

Chris Brantley, Manager, Government Activities and Operations, IEEE-USA

ciation, and the Christian Coalition. What do they all have in common? A very large and motivated membership that participates in grass-roots political activity on issues of common concern.

Equally enlightening was the respondents' list of what works in lobbying:

1. Delivering the straight facts to lawmakers;
2. Having active allies in a Congressman's district;
3. Mobilizing grass-roots action, such as phone calls and letters;
4. Getting along well with politicians and their staffs.

And what doesn't work? Johnny-Come-Lately's who rush into action during election years and spend their money on high-priced lobbyists and TV/radio/print attack ads.

During my nine years with IEEE-USA, I have frequently interacted with members who have seemed genuinely surprised with the notion that political popularity can dictate the legislative solution to a problem rather than a cold, hard systems analysis of the options, which could identify the most efficient solution. But to quote from Marc Caplan's *A Citizen's Guide to Lobbying*:

> Few bills, if any, pass solely on their merits. Evidence is nice, but facts do not vote; constituents do. Organizing citizen support for a bill is the most crucial part of lobbying.

Therefore, it isn't enough for engineers to offer the technically sound "right answer" to solve national, state, or local problems; they have to help ensure that the decisionmaker who is being asked to implement that solution is comfortable that it has broad public support and will help, not hinder, re-election prospects.

Hopefully, these excerpts, combined with the exhortations of my fellow panelists, have convinced you of the critical role that grass-roots activism can play in shaping our laws and regulations at the federal and state levels. But if you are still in doubt, I leave you with this advice given by Congressman George Brown to IEEE-USA at its 1997 Technology Policy Symposium:

> You also need to use your membership to engage in a broader process of educating legislators and the public. Professional science and engineering societies should be using their local chapters and regional sections to interact with members of the House and Senate. These members should be helped to realize that these seemingly arcane debates about technology development have a local face at high-technology companies in their states or districts, or at colleges and universities at home. They need to gain a better understanding of your world and the realities of our science and technology efforts.

WHY ME?

Another easy answer—self-interest.

It must have happened to you. One day you opened the paper or turned on the TV and saw something that put a crease in your brow or caused your stomach to churn. You scratched your head and said "that doesn't make sense" or "that can't work." Or you realized that this proposal or the failure to take action (i.e., political gridlock) could adversely affect your job, hurt your community, or put a strain on your bank account.

So that my point is not too abstract, here are some concrete examples. It could be that proposal being considered by your state board of licensing and registration to impose mandatory continuing education requirements. Maybe you're a software engineer in Texas, and you are subject to a new registration requirement. Perhaps you're nearing retirement and thinking about that change in the pension laws under consideration in Washington that could improve your retirement security, if only Congress would move it forward. It could be that your company let you go and you just heard that a foreign engineer brought into the country with an H-1B guestworker visa took your place. Or your research proposal was rejected because of federal R&D fundings cuts. To close out this parade of horribles, perhaps your state is planning to deregulate your electrical utility service and you're fearful that their proposal will only serve to increase the cost of electricity, add to your frustration, and maybe even damage the reliability of your electrical service.

You'd like to protect your interests and encourage the best possible solution to the problems, but what do you do? What can you do? And if not you, then who?

The first thing is to become politically active as an individual. That means becoming more educated about the issues and communicating your views to your elected and appointed representatives. It means establishing a relationship with them, if possible, by visiting them, becoming a credible resource for advice and information, inviting them to visit your company, university, organization or community group, and volunteering to work on their political campaign. Last, but not least, it means joining with and recruiting others who share your concerns and interests to do the same.

HOW TO BUILD A LOCAL GRASS-ROOTS NETWORK

A grass-roots network is essentially a group of individuals with similar interests and concerns who are willing to take personal action. So where do you start? Why not right in own your local IEEE section or group, in your church group, or with your neighbors.

Begin your network by creating a list of everyone you know who might be able and interested in helping. Then contact them and ask them to volunteer. Build a phone tree or e-mail list so that you have a way to contact everyone who shares your interest. Share the burden and get everyone involved. As your network grows, recruit coordinators (i.e., individuals who are willing to take responsibility for keeping a portion of the network informed and active). Ask your network members to recruit new members—those who are successful will become the coordinators.

Use technology to help manage the network. Database or address book software can be used to keep contact information and generate mailing labels. E-mail provides a cheap and quick means of communication within the network. You can save the costs of postage and minimize e-mail traffic by providing background information such as draft legislation, proposed statements, and points of contact on the World Wide Web.

Once your network is established, it is also important to broaden your base by building coalitions with other networks and organizations willing to invest their energy and resources on issues of common concern. If it is a technical issue of interest to IEEE members, it may affect certain other engineering disciplines. Investigate what other societies and their local sections are doing. If your issue concerns your community or state, any number of organizations and groups may be willing to join forces. You may have to educate them on the issues, but once they understand how they can be affected, they will look for ways to endorse and support your efforts.

For years, IEEE-USA has lobbied in Washington for greater portability of pensions. Pension portability is the perfect example of an issue that can cut across a broad spectrum of groups and professions because of our increasingly mobile U.S. work force. Accordingly, IEEE-USA's pension lobbying efforts are coordinated with other organizations through the Pensions Coalition and the Engineers and Scientists Joint Committee on Pensions.

The downside of coalitions is that it can be difficult to reach consensus on priorities and lobbying strategies when there are more players at the table. However, it is usually better to approach decisionmakers with a coordinated voice and to leverage the available resources for a common purpose, rather than generating confusion and possibly canceling out each other's influence with multiple and conflicting efforts.

THE TOOLS OF THE TRADE

So you've got a grass-roots network in place. How can you make effective use of it? Here are some notes on grass-roots lobbying, including some tips on useful tools of the trade:

1. Energizing Your Network.
One of the real challenges in grass-roots lobbying is how to get volunteers who have expressed an interest to take action when needed. Paul Revere's midnight ride would have been to no avail if the Minutemen had not grabbed their muskets and marched to meet the British regulars. Having a list of volunteer names in a database is not enough; you need to keep that network energized so that they are ready to act when the alert goes out.

The basic formula for keeping a grass-roots network energized includes three steps: education, motivation, and activation. *Education* involves keeping your network members up to date on the issues and developments. Make sure they understand the politics and the key players who are also attempting to influence outcomes on your issues. Give them advance notice of decision points, such as hearings and key votes, and warn them that they will be called upon to take action. Grass-roots volunteers (especially engineers) must feel comfortable that they know what they are talking about before they will take action. Legislative Bulletins and/or newsletters are typical tools for keeping your grass-roots network up to speed. Some other tools include regular meetings, e-mail lists, and the World Wide Web.

Motivation is also a part of education. It is not enough to tell network members to write to their legislator to urge a particular action. The grass-roots volunteers also need to know how the legislation will affect them, their interests, or their community. Will it affect their pocketbook (and how)? Will it affect how they practice their profession (and how)? Will it somehow hurt the community, state, or country (and how)? Unless your volunteers are motivated by personal interest and knowledge, they will not respond to your call for action.

Finally, there is *activation*. Typically this is done by issuing an Action Alert to your network members, asking them to take a specific action by a certain date. This alert gives them necessary contact information, explains why the action is necessary, and provides talking points or background information to help them respond. Members should be activated at those key points in the legislative process when their input can make a difference. Contacts should be scheduled to coincide with key steps in the legislative process, including bill introduction, hearings, legislative mark-ups, subcommittee and committee votes, floor debates and votes, and conference committee deliberations. If the bill is not moving, then action alerts can be used to urge a hearing or encourage co-sponsorship of the bill to move it into the legislatie spotlight. Timing is everything. If input is provided too early, policymakers will ignore or forget it by the time they have to take action. Another common mistake is to issue the call for action so late that network members either don't have time to respond or their responses are received too late, which conveys the unfortunate message that your volunteers are not politically savvy.

Like any tool, a grass-roots network has to be used regularly to stay sharp. People need to be involved and feel that their involvement is making a difference; otherwise, their interest will fade. You can't put your network on hold after the legislative session and expect that people will respond quickly when the next session starts. By the same token, activation can't be just busy work; it has to make your grass-roots volunteer feel like he or she is making a real contribution. And you have to stop to acknowledge and celebrate your successes, one of the most common oversights in our sometimes frenetic existence.

2. Contacting Your Legislator.

Once you've energized your grass-roots network to take action, there are three basic tools that your network members can use to reach a legislator or government official quickly. I'll limit my comments on each to a few "how-to" highlights, but I encourage you to read IEEE-USA's "Engineers Guide to Influencing Public Policy" on-line at *http://www.ieeeusa.org/usab/DOCUMENTS/FORUM/LIBRARY/GUIDE* for more detailed suggestions.

Phone calls are the quickest method to register your views. You can find your Senators' and Representatives' phone numbers in the blue pages of your local phone book or by contacting the IEEE-USA Washington office. Don't expect to speak with the actual member of Congress unless you are well-known to each other, but ask to speak with the staffer who is responsible for monitoring the issue or bill of concern to you. When you get that person on the phone, identify yourself as a constituent and then tell them what you want your member of Congress to do and why. Keep your call short and to the point; don't threaten or talk down to the staffer. Offer to provide more detailed information in support of your position, if necessary.

Letters can be effective, if you identify yourself clearly as a constituent, if the content is clear and to the point, and if you focus on one topic and state clearly what action you would like taken and why. Personal letters carry disproportionate weight, since members assume anyone who is motivated enough to take the time required to write such a letter represents the tip of the iceberg of constituent concern. On complex science and technology issues, IEEE-USA has been told frequently by our contacts on the Hill that as few as four to seven personal letters from constitutents are enough to encourage that member to take action. Form letters, post cards, and blast faxes on a specific bill or issue are counted and weighed and may be influential in large quantity, but are typically not read or responded to.

E-mail is used more and more for grass-roots communications and by members of Congress and their staff, although acceptance is still not universal. Many offices routinely delete any e-mail that is not obviously from a constitutent, so make sure your e-mail has a mailing address or advisory in the subject heading and at the top of the page. Only a few offices review e-mail in its electronic form; many just print off the messages and add them to the stack of constituent mail to be reviewed and answered (or not). In that case, the same advice for writing an effective letter also applies to e-mail.

In addition to a quick-response mechanism, an effective grass-roots network should also be involved in supporting a more deliberate, proactive plan for interaction with your elected representatives. The goal is to establish an on-going relationship with the legislator so that he or she knows your group and interests and knows whom to contact for information and support. Activities and events such as visits to a legislator's Washington or district office, invitations to speak at meetings and special events, demonstration and tours, award presentations, etc., are all key components of an effective plan.

3. Working the Media.

One important way to influence legislation or policy is to raise public awareness and sway public opinion on your issues. Doing this involves working with the media and establishing yourself (or your network or coalition) as a credible source of information that can respond to reporters' requests on short notice. There is no substitute here for establishing a working relationship with the reporters who cover the issues you're concerned about. You can do that by attending the events that they cover and talking to them, sharing useful information, and identifying credible sources that they can interview. If you help them do their job well, they find a way to repay the favor.

Press releases are the traditional tools used to convey your message to media outlets. Whether your message will see print depends largely on several factors, including how "hot" the issue is, how newsworthy your contribution is, how credible your information source or organization is, and how well drafted the release is.

Letters to the Editor and OpEds also get your message into print. When you see an article in the local paper on an issue of concern and you think the article missed a key point or got something wrong, don't wait for someone else to say so (since they probably won't). Take the time to write a letter to the editor and submit it as soon as possible so that your response is timely. OpEds are prepared editorials on certain subjects that are submitted for print. It is worth the time to contact your editorial page editor to find out what they expect in an OpEd piece. If you can also arrange for a prominent member of the community to submit the OpEd, it will increase the prospect that it will be printed.

Television and print interviews also raise public awareness. When the opportunity for an interview arises, it is critical that your network/coalition put forward the best possible speaker—someone who has a good command of the issues, communicates well, is able to speak in quotable "sound bites," and will stick to the network/coalition message. This can be difficult, since media opportunities typically arise on short notice. It is best to recruit a spokesperson well in advance and if possible to practice for media opportunities.

Another way to get media exposure is by showing up at press conferences and media events with banners, signs, and spokespeople who can share your group's perspectives with reporters.

For more tips on developing an effective public relations effort, consult Chris Currie's advice on implementing a section media strategy at *http://www.ieeeusa.org/usab/PACE/LIBRARY/media.html*.

4. Hiring a Lobbyist.

A lobbyist is simply someone paid to help influence legislation or regulations in a way that serves the interests of the client. A good lobbyist is someone who is knowledgeable about the issues and the policy process, who has a network of contacts that can be used to build coalitions and access policymakers, has experience and a successful track record of advocacy, and is not subject to conflicts of interest. If funds are available, a lobbyist can provide valuable assistance to a grass-roots network by helping to monitor legislation and the legislative process, generating a lobbying strategy for your network, shaping your message and identifying opportunities to communicate it, and serving as a primary point of contact for the network, media, decisionmakers and others.

Hiring a lobbyist is not an inexpensive proposition, but if you are able to do so, then you will need to negotiate a contract that specifies in some detail the duties of the lobbyist and fees to be paid. Your lobbyist should also be registered with the appropriate state agency and/or U.S. Congress.

Some engineering societies provide a modest stipend or expense allowance to a retired member who is willing to take on the job of being a lobbyist. This is fine as long as that individual satisfies the applicable lobbyist registration requirements. Your network may also be able to draw on the lobbying experience of the organizations to which the network members belong. IEEE-USA for example, has five professional staffers who are registered as lobbyists, who are available as a resource to assist IEEE's U.S. members.

One other point about lobbyists is that if you are hiring a lobbyist or engaging in lobbying activities as an IEEE entity, you need to report your activities and expenditures to IEEE's Manager of Tax Compliance and to IEEE-USA so that we can file the proper IEEE disclosure and financial reports with the Internal Revenue Service and the U.S. Congress.

5. Getting Involved in Campaigns.

There are three basic ways you can get involved in campaigns: fund-raising, candidate endorsements, and campaign volunteering.

One prevailing notion of lobbying is that in order to achieve your legislative goals, you have to be willing to raise money for political candidates, typically through political action committees or by personal gifts. Money helps candidates get elected and therefore your campaign contribution buys access and influence with the newly elected decisionmaker.

It is certainly true that campaigns turn on money. The cost of a typical campaign for a U.S. House of Representatives seat is now estimated at more than a million dollars. Candidates for the Senate may spend $5 million to $10 million for a successful campaign. But remember what campaign money buys—exposure of the candidate to the voters through mailings, TV time, travel, etc. If you can deliver the voters through your grass-roots network, you don't need to contribute money to the campaign.

Campaign volunteering can take any number of forms. Typical activities might include staking a candidate sign in your front yard as a personal endorsement, going door to door to distribute information about the candidate, organizing and hosting a fund-raising event, helping to develop campaign positions on issues that tap your own expertise, or working in the campaign office stuffing envelopes, making cold calls to voters, or whatever else needs to get done.

As a 501(c)(3) non-profit society, IEEE is not allowed to participate in partisan politics. Among the proscribed activities are participation in political campaigns, gifts to candidates, and candidate endorsements. Therefore, as individuals or networks working to advance IEEE's legislative agenda, you should not engage in any of these activities in IEEE's name or using IEEE's funds. However, you are free to get personally involved in a campaign and contribute your own funds if you wish to do so; just make sure that IEEE is not explicitly or implicitly connected with the activity.

WHAT IEEE-USA IS DOING ABOUT GRASS ROOTS

The importance of having a strong grass-roots network for public policy advocacy is not news to IEEE-USA; we've tried a variety of mechanisms over the years for reaching our members and getting them involved in grass-roots advocacy with varying degrees of success.

When I joined the IEEE-USA staff in 1989, we maintained a key contact database of names, addresses and interest profiles that could be used to generate mailing labels for Legislative Alert mailings. However, tight budgets, timeliness concerns, and the small size of this database (1,200 names at its peak) discouraged its regular use, and it never became an effective tool.

Later, we worked with the member database operators at IEEE's Operations Center in New Jersey to install sorting software that would allow us to generate mailing labels by Congressional District using member zip codes. This proved to be of limited use because most zip codes in the member records were only five digits, which allowed for an accuracy of approximately 50 percent when matching against congressional districts. More recently, IEEE has been able to upgrade the zip codes to nine digits, using software provided by the U.S. Postal Service. With nine digits we can now generate mailing labels by congressional district with approximately 95 percent accuracy. However, the old problem of mass mailing postal costs has limited our use of this tool.

In recent years, instead of focusing on building a separate grass-roots network, our emphasis has been on reaching out by mail or e-mail through existing networks. Depending on the issue and the time available to respond, we have used PACE, the IEEE-USA all-volunteers list, specific committee mailing lists, the consultants networks, the IEEE-USA Board of Directors, selected IEEE mail lists, or some combination of all of the above to circulate information bulletins and action alerts on fast-moving legislation.

As the ability to generate constituent input to Congress has become more important, however, it is also increasingly clear that our previous methods are not sufficient and that a true grass-roots advocacy network is needed. At the 1997 Professional Activities Conference, Mark Pullen, Vice Chair of IEEE-USA's Technology Policy Council and a former Congressional Fellow, made a strong case for the necessity of having an effective network. Based on the response he received there, IEEE-USA experimented with a small electronic network of PACE volunteers in 1997, but discovered that the response rate to requests for assistance was disappointingly low. Staff and budget support for this experiment was also minimal.

This experiment underscored what is a daunting question for managers of grass-roots networks: how can we recruit and retain members who will respond when asked? Based on an examination of successful organizations and networks, the apparent answer involves:

- Focusing on issues that are of personal interest to the participating members and making sure they understand how these issues can affect them personally so that they have a strong motivation to respond;
- Providing participants with the right type and amount of information on the issues and in the right format so that they are informed but not overwhelmed by the communications;
- Communicating information with the right frequency so that members are aware and comfortable when asked to take action and are not asked to respond to requests "out of the blue";

- Making participation as easy as possible by using technology where possible, but always maintaining a human element; and
- Showing network members how their participation helped make a difference so that they feel that their efforts are worthwhile.

Translating this into a real-world plan means changing priorities, increasing staff and volunteer time spent on grass-roots advocacy, making sure our efforts are funded adequately, and applying technology to leverage our resources.

As this article was going to press, IEEE-USA's Board of Directors was considering a 1998–2000 Strategic Plan that included a priority initiative for a pilot project to establish a network for federal, state, and local grass-roots advocacy. IEEE-USA's government relations staff has been charged to develop an action plan that includes success metrics and milestones for implementing this project. The goal is to develop a network capable of delivering thousands of timely letters and e-mail to Capitol Hill on priority issues and critical legislation.

Here are some of the assumptions we are making as part of our preliminary planning:

1. The network should be electronic (as opposed to mail- or fax-based) for budgetary and timeliness reasons. We should make use of one-way majordomo e-mail lists with cross references to legislative and advocacy information stored on the World Wide Web. We should use the World Wide Web as an interface to sign up members interested in grass-roots advocacy and as a one-stop source for advocacy information.

2. The network should be organized so that we can build majordomo e-mail lists targeted at key congressional districts and also at our highest priority issues (our legislative agenda). It is probably not realistic to establish lists for all 435 congressional districts, but we will start by identifying priority districts that are geographically distributed. For those districts that don't have a list, we'll invite local advocates to recruit a specified number of interested members in that district and then develop a majordomo mail list for our/their use once we reach that threshold.

3. We need to investigate what technologies are available to help us build a viable electronic network. Two possibilities for exploration include on-line data collection so that we can construct a database of participant information (e.g., congressional districts, policy interests, key contacts) and an interactive forum that allows members to identify their congresspeople by zip code and send letters online. Part of the investigation should include an examination of whether we can apply push technologies to inform and involve members according to their personalized interest profiles.

4. Staff responsible for specific issues will have to assume responsibility for maintaining the issue-oriented advocacy lists/networks. Each of our staff government relations professionals will have to integrate grass-roots advocacy into their respective

functions to make sure that we can develop and provide the right information content with the right timing to our network participants. This will require setting priorities and refocusing our efforts, which cannot be achieved without volunteer support.

5. We would need to work out a plan for managing our district-specific lists. There would also have to be some overall coordination and high-level planning.

6. The PACE Network can play an extremely important outreach role in helping us to publicize and recruit members to the network(s).

Based on these assumptions, a number of issues will also have to be addressed as we move an action plan forward for developing a grass-roots network. We will need a "marketing" strategy for recruiting network members, agreement on preferred practices for utilization of the network (e.g., how to package information "content," how often to send out information to keep participants up to date and motivated without overwhelming them), metrics for success, and milestones for implementation. We'll have to examine how the recently issued IEEE e-mail policy affects our ability to recruit and communicate with network members by e-mail. And we have to think about how to complement or integrate existing networks within PACE and at the regional, state, and section levels. Finally, once we have worked out answers to these questions, we have to get the IEEE-USA's Board of Directors to commit to an implementation plan, with goal(s), milestones, metrics, budget requirements, and staffing implications.

CONCLUSION

You have a personal interest in speaking out on the issues and bills that will affect you, your career, your family, and your community. And with the tools and methods described here, you can recruit others with similar interests and concerns so that one voice becomes many. Your elected representatives will listen to you if you take the time to approach them in the most effective way. Do it for yourself, not necessarily for IEEE-USA. But remember you are not alone; we are here as a resource for legislative information and advocacy tips that will help you help yourself. Just contact me in the IEEE-USA office and/or visit our Public Policy Forum on the World Wide Web at *http://www.ieeeusa.org*.

ABOUT THE AUTHOR

Chris J. Brantley is manager of government activities and operations for the Institute of Electrical and Electronics Engineers-United States of America. A registered lobbyist, he is responsible for IEEE-USA's government fellowships and internships, state government activities, and implementation of a new IEEE-USA Grass-Roots Network. Prior to joining IEEE in 1989, he was assistant to the executive director and acting director of government relations for the American Association of Engineering Societies. A complete resume is available on-line at http://www.erols.com/brant/brantley.html. He can be reached by e-mail at c.brantley@ieee.org.

Recommendations for Effective Communications to Influence Federal Policy Issues of Importance to Electrical Engineers
R. M. PAUL

ABSTRACT

Engineers have a compelling story to tell. If they can tell it in a form policymakers understand, those leaders will give it greater consideration. The challenge lies in putting a "human" or "emotional" face onto highly technical issues; most policymakers have limited technical background and little time and resources to learn about such issues. By communicating effectively, engineers create understanding and "buy-in" from federal policymakers while promoting the engineering profession. Engineers can learn to communicate at the "grassroots" or local level; such effort has proven effective in influencing federal policy decisions. This paper offers recommendations for becoming more conversant and effective in communicating with your federally elected representatives.

WHY SHOULD I CARE?

Federal policy decisions made in science, engineering, and technology areas have an impact on every profession. This is true whether you work in industry, academia, or as a consultant. Policy decisions have an impact on the direction and resources that drive research and development in the public and private sectors. One of the most influential factors affecting federal policymakers' decisions is what they hear from home—grassroots communication. This is where your input to your Congressman and Senators can play a pivotal role.

If you are working with the impression that you can pursue your profession without staying informed of policy decisions and political trends, consider Microsoft founder Bill Gates' recent opinion on the matter: "We need to increase our dialogue with political leaders so they understand the excellence we stand for."

The highly technical nature of engineering requires a more concerted effort from engineers to establish communications and dialogue with policymakers. Such effort will

---

Raymond M. Paul, Administrator, IEEE-USA Technology Policy Activities

help ensure that our federal leaders, who, for the most part, do not have technical backgrounds, give proper consideration to and make well-informed decisions on technical issues. Rest assured, policy decisions that will affect your career will be made regardless of whether they are informed decisions or not. I made both in my career as a Senate staff member. Don't you think, then, that it is in your best interest to do everything possible to be sure the decisions being made *are* informed ones?

YOU HAVE DECIDED TO BECOME INVOLVED IN THE FEDERAL POLICY PROCESS. NOW WHAT?

Having served in both the public and private sector for a number of years, I have been on both sides of communications campaigns. Some of the communications strategies used were successful, while others were less effective. Communicating effectively and presenting your ideas and knowledge are essential to a successful career. It is also your privilege and right as a U.S. citizen to do the same with your elected officials. After you develop and deliver an effective message, your priority will be to remain in contact with policymakers (usually their staff) and to offer further assistance.

LEARNING TO TALK THE TALK

To communicate effectively, you must understand your audience. This may seem simple at first. However, engineers and policymakers generally speak two different languages and have different thought processes. Keep in mind that policymakers and their staff will (and often do) make technology policy decisions with little technology expertise. Therefore, it falls to engineers to learn some of the "policy talk," and more importantly to learn how to communicate recommendations effectively to an audience that is very different from your profession's typical audience.

One of the first steps toward communicating effectively to an audience is to feel comfortable and in control yourself. When you know what you want to say and deliver your message well, your audience will be more inclined to pay attention to you.

CONTENT PREPARATION AND ORGANIZING YOUR THOUGHTS: BUILD A STORY

"Making Perception Reality"

It is an unavoidable truth that the policymakers' perceptions generally carry at least as much weight in the decision-making process as facts and political factors. In fact, one of the largest and most successful public affairs companies in Washington, D.C. states this notion on its letterhead. The growth and implementation of this notion as a tool leads to a "chicken-or-the-egg" question to explain the explosion of the use of grassroots networks to influence policy. If policymakers hear from a large number of constituents on an issue, they inherit the perception that a majority of their voters feel the same and therefore they should support the popular position. In reality, grassroots activity is a way of gaming the system, but it is very effective.

The greatest hurdle to overcome is the reality that on the surface, engineering doesn't always present a story that inspires people who lack a technology background. Basically, engineering lacks a human or emotional factor that will engage non-engineers—your audience.

Because of this, you need to develop a message that strives to do the following:

- *Inform*—Provide the necessary facts and appropriate details. Your message should contain only the facts necessary to make your point. Don't overload your message with facts; you are not developing a technical paper for a technical audience.
- *Motivate*—Make recommendations that will allow your audience to follow a specific course of action. Offer specific ideas about how to improve specific pieces of legislation or existing policy.
- *Persuade*—Convince your audience that your message is valuable. Use anecdotes to "bring home" the issue and put a human or emotional face on it. Remember your audience; empirical data can and does play a role in persuasion, but policies are made by buyers (Congress/Administration) and sellers (constituents/lobbyist), whose market primarily consists of ideas.

Many techniques exist to help you achieve these communication objectives. These techniques and tips include the following:

- Keep your language non-technical to the greatest extent possible. Do not make your audience members feel like children, but remember, if you use terms that they do not understand, your message will be lost.
- Draw on appropriate personal experiences and illustrate them to support your message, but remember:
  —Relate that experience to the audience. Finding common ground between you and your audience can help bridge differences.
  —Avoid expressing your position as a personal opinion. Opinion gives your audience the choice of agreeing or disagreeing with you. You stand a better chance of persuading them to accept your *position* on an issue.
  —Use analogies to clarify your point, especially if the issue is complex. Accessible analogies are imaginative and support your argument *quickly* and *clearly* in terms that are understandable to your audience.
  —Cite relevant applications of your ideas to which your audience can relate. Remember that you will need to bridge an understanding to technology. For example, many non-technical people have access and familiarity with global positioning satellite (GPS) because they use it while hiking, fishing, or sailing.

But few users have any idea how GPS works, other than "it has something to do with satellites." They have no appreciation or knowledge that it depends primarily on accurate clocks and that, the more accurate the clock, the more accurate the GPS reading will be. Find out if your elected official likes to fish; if he does, tell him how your profession is responsible for designing systems that help him return to his favorite fishing hole. Again, "humanize" or "bring home" your position.

—If you are having a meeting personally with an official or staff member, present a *concise* hypothetical case or scenario. A well-crafted hypothetical example that dramatizes your position can illustrate the potential impact—either positive or negative—a decision can have. "Show-and-tell" your audience what will happen if your position is not followed or your ideas are not implemented.

—Cite outside experts or interested third parties; objective testimony or evidence can lend credibility to your position. However, do this with caution; outside the science, engineering, and technology community, few people know the difference between a Vannevar Bush and a Kirchhoff law.

—Pose questions that you can answer positively to defuse potential adversary positions. This will add to your credibility and indicates to your audience that you have mastered the nuances of your subject matter.

CONCLUSION

The most important thing to remember in developing an effective message is that you must understand who your audience is and what creates their "buy-in" to ideas. Where the engineer thinks in terms of ordered systems that are governed by facts and research, the policymaker thinks more in terms of human emotional appeal, which is closer to the chaos theory. With this understanding, engineers can overcome the "language barrier" hurdle and communicate their ideas and positions more effectively to policymakers. I have outlined some of the effective techniques and strategies that I have found during my Capitol Hill and private-sector experience. I hope that you will find them helpful as you work to provide input on engineering and technology policy interests to your Congressman and Senators.

ABOUT THE AUTHOR

Raymond M. Paul is administrator of technology policy activities for the Institute of Electrical and Electronics Engineers-United States of America. A registered lobbyist, he is responsible for helping volunteers develop policy positions in the fields of energy, environment, defense, and aerospace. Prior to joining IEEE-USA in 1997, he was a lobbyist for Hill and Knowlton Public Affairs Worldwide, and he served for more than six years as a professional staff member for former Sen. Bennett Johnston of Louisiana.

Employers' Engineering Education Needs for the New Millennium

L. E. BRYANT

ABSTRACT

The Foundation Coalition received a grant from the National Science Foundation beginning in 1993 to focus on changing the ways in which engineers are educated to meet the needs of industry better. A National Advisory Board was established as part of this effort to provide guidance on addressing the desired qualifications for engineers employed by government and industry. A mission statement that addresses six qualities required for graduate engineers has been developed.

BACKGROUND

In 1993, the National Science Foundation (NSF) established the Foundation Coalition as the fifth group of education institutions to receive grants focusing on new methods for engineering education. These grants were established in response to feedback received from various groups, including industry leaders, the National Academy of Engineering, the Accreditation Board for Engineering and Technology, the American Society for Engineering Education's Engineering Deans' Council, the American Society for Engineers, and the National Research Council Board.

The seven founding institutions of the Foundation Coalition consisted of the University of Alabama, Arizona State University, Mesa Community College, Rose-Hulman Institute of Technology, Texas A&M University, Texas A&M University-Kingsville, and Texas Woman's University. Forces that brought these institutions together included the quality of their engineering education programs, diversity of the student population, willingness to dedicate time and resources to education change, industries served by the institutions, and/or geographic diversity. In addition to the educational entities, the Foundation Coalition established a National Advisory Board (NAB) made up of industry and government employers of engineers who were focusing on their engineering needs of the future and were willing to work with universities to bring about changes to address these needs.

LeEarl Bryant, P.E., Texas LAB Consultants

VISION FOR NEW ENGINEERS

Meeting twice a year with academic leaders, members of the NAB continually review the direction of the program, provide guidance for changes, and when possible, provide fiscal resources and/or co-op/summer jobs for program participants. During the years since the program began, the NAB has continued to confirm the need for a new generation of engineers and has worked as part of the Foundation Coalition to maintain focus on a strong engineering education with improved curricula and learning environments. These environments are structured to resemble the real world in which graduates will have to work.

Working with the NAB, the Foundation Coalition has developed a mission statement, which envisions engineers with an:

- Increased appreciation and motivation for life-long learning;
- Increased ability to participate in effective teams;
- Effective oral, written, graphical and visual communication skills;
- Improved ability to appropriately apply the fundamentals of mathematics and the sciences;
- Increased capability to integrate knowledge from different disciplines to define problems, develop and evaluate alternative solutions, and specify appropriate solutions; and
- Increased flexibility and competence in using modern technology effectively for analysis, design, and communication.

ABOUT THE AUTHOR

LeEarl Bryant has a broad-based background in engineering and management. She has been employed by large industry (Rockwell International), has participated in numerous ventures, and is now serving as an independent consultant in project management, business planning, and technical writing. Throughout her career she has been active with universities and professional organizations, with a focus on quality engineering education. She is currently serving as a member of the Foundation Coalition's National Advisory Board.

A registered professional engineer in the state of Texas, Ms. Bryant obtained her bachelor's degree in electrical engineering from Texas Tech University and her master's degree in electrical engineering from Southern Methodist University.

Developing and Maintaining a Competitive Career

K. BUCKNER

ABSTRACT

Career development does not have to be a mystery. When the organization provides a solid framework, a common language, and an effective set of tools, individuals can take responsibility for managing their careers in today's dynamic business environment. More than 20 years of research have provided a model that supports strategic individual development. This model, coupled with identification of career drivers and 360-degree competency-based feedback, give employees the information necessary to develop powerful individual development plans. The plans have a dual focus: (1) how to remain competitive through personal growth and development; and (2) how to increase their level of contribution to the organization.

Participants will learn:

- How to identify their individual genius and how to apply it to their work.
- How to map their own career drivers and determine how these drivers have an impact on how they feel about their job.
- What organizations expect of individuals and how those expectations change over time.
- How to use 360-degree competency-based feedback for creating effective individual development plans.
- How organizations have applied these tools to build organizational capability.

Participants will receive a copy of the research-based *Career Success Map*. This self-assessment instrument helps individuals identify their personal career drivers. They will also receive a copy of *Turning Feedback Into CHANGE!* by Joe Folkman, Ph.D., which teaches how to use feedback to bring about genuine and positive change in our own behavior. It offers practical, simple suggestions and entices us to start the process and follow it through.

Kathy Buckner, Novations Group, Inc.

INTRODUCTION

As the organizational competitive environment becomes increasingly complex, new demands have been placed on both individuals and the business in which they work. Most large companies have responded to increased external demands by increasing structural efficiency (downsizing) and trimming bureaucracy (flattening). This move to more austere structures has not been a matter of simple cost-cutting. Both global and domestic competition have intensified dramatically. These new demands call for a greater contribution on the part of all employees, and for greater collaboration within and between work groups.

Research indicates, however, that organizational culture, values, and systems have generally lagged behind this revolution. One example of this gap is that employees have been told that they are responsible for managing their own careers. However, many organizations have failed to equip people with the tools they need to succeed in this new duty. A number of important developments make this an increasingly important problem:

- In the past, the development process was focused largely on preparing employees for the next promotion. Upward momentum in the organization was seen as the primary indicator of good performance, and employees were motivated by the promise of moving up. More complex jobs automatically provided increased challenge; employees had to learn in order to succeed in their new roles. For most people, regular promotions are no longer feasible.
- People are a key source of competitive advantage. As structures have become leaner, the net value of each remaining employee has increased. Moreover, the new "knowledge economy" means that more and more companies rely on the education and experience of highly sophisticated knowledge workers. As the business environment changes, retooling often means updating the skills and abilities of the organization's people rather than changing over a plant. On-going development is a competitive necessity.
- In many fields, increased competition for capable and qualified employees means that the best people often choose employers based on perceived opportunities for growth. In an era when promotions are few and far between, many of today's technical experts look for cutting-edge learning opportunities, increased autonomy, or balanced lifestyles as acceptable (or even preferable) substitutes.

In spite of these dramatic changes, most career development systems are still based on assumptions that developed early in the industrial age. This article will explore alternatives to the traditional model and will provide examples of innovative replacements.

CAREER BESTS: WHERE DEVELOPMENT OCCURS

Employees and the companies for which they work each have an interest in the employee's long-term growth. Development opportunities (often framed in terms of increasing "employability") are offered up as a modern-day replacement for long-term employment; smart employees know that the key to long-term success is now grounded more in maintaining sharp skills than in demonstrating loyalty to the company. Development also benefits the company because it leads to the appreciation of human assets.

Research on what makes a satisfying work life indicates that the most personally satisfying times in a person's career are usually also highly productive in terms of meeting the organization's goals. Data on such "career bests" indicate that peak experiences also provide significant development. One of the most frequently cited characteristics of a career best experience is challenge or learning opportunities. Rising to the development challenge benefits employees by building increased capability and satisfaction. Increased individual capability in turn increases the organization's ability to perform. This concept is illustrated in the diagram below:

Career Bests

Career bests almost always happen when individuals are doing something that they enjoy, that uses their talents, and that falls in the domain of strategic business needs. In other words, long-term career success results when people identify the shaded area in the diagram above and figure out a way to spend as much time as possible there. The fact that the best development occurs when the individual and corporate interests intersect or overlap has several important implications:

- It underscores the importance of individual responsibility for career development. Individual interests vary from person to person, and only the individual knows where his interests lie. Finding the career best zone and staying there can't be done with a career plan developed by someone else, whether it be a manager, mentor, or human resource department.
- The best development plans do not center on next jobs or suggested training courses. While new job opportunities can provide growth experiences, and while training courses can augment on-the-job learning, most development happens as the result of engaging in challenging, interesting work. That type of work can be found in most professional job assignments and rarely requires a job change.
- The organization must be clear about what it needs from employees. Rapid changes frustrate many employees, who feel as though they must constantly shift their aim in

order to hit the moving target of organization direction or expectations. This problem is likely to continue. However, leaders can share information about the organization's direction and can translate that direction into individual expectations. One common approach for doing this is to identify competencies, which reflect the organization's priorities for individual performance and development.

• Self-directed career development requires more self-awareness and insight than corporate-driven career management. Individuals must take more initiative and responsibility to understand and articulate their own needs, priorities, and ability to contribute than they did in the past. This may be a painful process for some, but is a liberating process for others.

IDENTIFYING INDIVIDUAL INTERESTS

One framework for understanding the individual's needs is to answer two key questions:

1. What can I contribute?

Most professionals work for more than just a paycheck. While monetary rewards are the most easily measured incentives, career choices are often based on other factors. Most people want to contribute something meaningful to a worthwhile purpose. Identifying individual priorities is a complex process because definitions of what is meaningful and worthwhile vary from person to person.

One tool for identifying how individuals can contribute is the concept of *individual genius*. In planning their development, individuals cannot ignore certain organization realities: constant change, increased demands, and reduced loyalty. These realities can be managed more effectively if they are weighed in relation to who a person is, what she likes to do, and what she does well. Those who take organizational changes in stride (and who capitalize on the opportunities that come with change) are those who understand and are true to what is right for them. Those who make the most significant contribution over time are those who know how their unique qualities add value to the organization. While understanding one's own genius or one's "personal truth" may be a life-long process, the genius concept can be broken down into two fundamental parts:

> *Interests/Passions*
> +
> *Talents/Abilities*
> =
> *Individual Genius*

Understanding one's genius is a key to increasing one's contribution. Some people make career choices based only on their aptitudes. They usually find that their ability or willingness to contribute over the long term is limited if they lack enthusiasm for their assignments. Doing what you care about automatically increases motivation, which increases the pleasure that comes from work. At the same time, all the passion in the world usually can't compensate for an innate lack of ability. The marriage of talent and passion leads to high performance. People who understand their genius and find ways to apply it in the organizations where they work build long-term value for themselves and for the company.

2. What are my values and priorities when it comes to work and career?

In the past, most employees set their sights on a fatter salary and a key to the executive washroom. In recent years, a much broader range of career drivers has emerged. One model that explains this emerging diversity was developed by C. Brooklyn Derr,[1] who identified five major definitions of career success:

1. *Getting Ahead.* This is the traditional definition of career success for most Americans. These people are looking to become vice presidents, presidents, CEOs, and general managers. What they want is upward movement. Success means more money, more power, and steady promotions—clear to the top.

2. *Getting Secure.* These people, an unappreciated and largely unacknowledged but significant segment of the job force, have a psychological contract with the company. In exchange for hard work and unswerving loyalty, they get life-long employment, respect, steady advancement, and eventually a high-level job where their talents are used and appreciated. They want to be a member of an organizational family.

3. *Getting Free.* These people want personal autonomy and "space" at all costs. They don't mind being held to deadlines, budgets, and standards, but they do want to solve the problem in their own way.

4. *Getting High.* These individuals thrive on excitement, challenge, and the technical nature and content of the work. They'll work for anybody who offers exciting opportunities—money is secondary.

5. *Getting Balanced.* These people give equal time and attention to careers, relationships and self-development. They'll work around the clock in emergencies, and they're happy to pay their dues, but they don't live their lives emergency-style. They usually pull back from getting overly absorbed in their work but are competent enough to do well at their jobs. Although they know how to negotiate and take time from work for themselves and their families, they are unhappy if their work isn't meaningful enough to balance their personal lives.

For most people, a career is more than a job. It is more than a long-term sequence of

[1]Derr, C. Brooklyn. *Managing the New Careerists.* Jossey-Bass, 1986.

jobs. Those who achieve career success acknowledge and respect aspects of personal life and of their own values that have an impact on work life. Effective career development allows people to live out the subjective and personal values they really believe in while at the same time making an effective contribution at work.

Since different people define their career goals differently, it stands to reason that they would need to employ different methods to achieve these goals. For example, if you want to get to the top of the organization, you will need to use strategies that are different from those used by the person whose primary focus is autonomy. When career values and the organization's needs do not match up, frustration and low performance result.

This provides yet another argument for why self-directed career development is the most effective approach. Individuals are much more keenly aware of their own career drivers and values than anyone else. When they use an understanding of those values to make decisions about their future, they are taking responsibility for their own development.

IDENTIFYING THE ORGANIZATION'S NEEDS

Increased competition has made clear that the expectations that most organizations have of their employees are much higher now than in the past. Standards of high performance are both more important and often less clear than they used to be. At the same time, the traditional measure of good performance—job promotion—is no longer a viable gauge. Although employees have been told to grow in their current jobs instead of focusing on the next one, most people do not have an alternative way of conceptualizing or discussing development outside the framework of job promotion.

One effective alternative to encourage development without focusing on promotions is the Four Stages$_{SM}$ Model developed by Gene Dalton and Paul Thompson.[2] While business school professors at Harvard University (and later Brigham Young University), Dalton and Thompson were asked to address a dilemma uncovered by the management of a large electronics firm. Their data (based on engineers' performance reviews tracked over time) made clear that expectations of individual performance change as people move through their careers. While some engineers continued to be rated as high performers throughout their careers, the majority received progressively lower ratings, even though the work they did remained the same in absolute terms. As Dalton and Thompson gathered data about what makes the difference between high performers and average contributors, the researchers identified four stages of development.

The progression identified by Dalton and Thompson is independent of position on the organization chart, and explains why two people with the same job description may

[2]Dalton, Gene W. and Paul H. Thompson. *Novations: Strategies for Career Management*. Scott, Foresman and Company, 1986.

be valued very differently by the organization. However, achieving high performance in the later stages depends on mastering the early stages. Thus, the Stages model provides a road map for understanding the long-term expectations organizations have of their employees. These expectations are described by the stages. Key tasks for each of the stages are summarized in the diagram below:

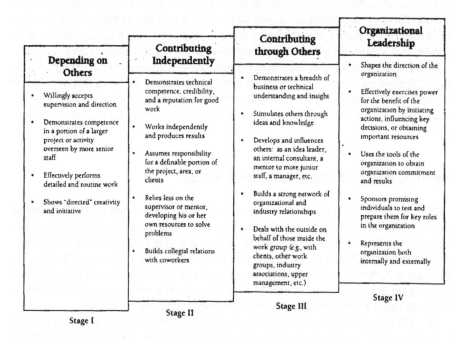

Depending on Others	Contributing Independently	Contributing through Others	Organizational Leadership
• Willingly accepts supervision and direction	• Demonstrates technical competence, credibility, and a reputation for good work	• Demonstrates a breadth of business or technical understanding and insight	• Shapes the direction of the organization
• Demonstrates competence in a portion of a larger project or activity overseen by more senior staff	• Works independently and produces results	• Stimulates others through ideas and knowledge	• Effectively exercises power for the benefit of the organization by initiating actions, influencing key decisions, or obtaining important resources
• Effectively performs detailed and routine work	• Assumes responsibility for a definable portion of the project, area, or clients	• Develops and influences others: as an idea leader, an internal consultant, a mentor to more junior staff, a manager, etc.	• Uses the tools of the organization to obtain organization commitment and results
• Shows "directed" creativity and initiative	• Relies less on the supervisor or mentor, developing his or her own resources to solve problems	• Builds a strong network of organizational and industry relationships	• Sponsors promising individuals to test and prepare them for key roles in the organization
	• Builds collegial relations with coworkers	• Deals with the outside on behalf of those inside the work group (e.g., with clients, other work groups, industry associations, upper management, etc.)	• Represents the organization both internally and externally
Stage I	Stage II	Stage III	Stage IV

Stage I contributors accept direction, establish basic competence, and learn the organizational and technical ropes. While people who do this early in their careers are seen as being highly effective, those who continue to depend on others for direction after several years are seen as contributing less than their peers who enter Stage II by becoming independent technical experts. By the same token, independent experts who fail to broaden their perspectives and develop others will be perceived as less valuable, unless they are brilliant enough to contribute as much on their own as their Stage III colleagues do by contributing through others. Those who are able to develop the Stage IV characteristics of having the vision and credibility to influence organization direction are perceived as being the most valued by leaders in the organization.

While the role changes suggested by the Four Stages_SM Model seem intuitively obvious on the basis of the traditional technical-to-management career progression, Dalton and Thompson found evidence that contribution, as described by stage, is relatively independent of the organizational hierarchy.

For example, Stage III characteristics (developing others, broadening perspective, understanding business issues, using networks to get things done) sound like a job description for supervisors or managers. Yet many people report experiences with supervisors who had formal management authority but few Stage III skills. On the other hand, one of the surprises from the research was that most of the people who perform Stage III functions are not in formal management positions: for every manager in Stage III, there were three non-managers performing similar leadership roles. While the proportion of managers to non-managers was reversed in Stage IV, there were still people in many organizations who guided the company's direction from their position on the technical bench, rather than from the executive suite. Recent research by Novations Group, Inc. suggests that the trend toward a higher proportion of non-manager Stage III and Stage IV contributors has continued with the advent of downsizing and flattening.[3]

The Four Stages$_{SM}$ Model describes values that have existed in large companies for decades. However, understanding the Stages Model has become extremely important as the size and shape of organizations has changed. Moving from one stage to the next increases an individual's ability to contribute—changing one's stage, or making a *novation*, can be done without changing jobs.

The word "novation" is a legal term meaning *the renegotiation of roles and responsibilities by parties to a contract*. A legal novation is a very formal process; career novations are usually much less formal. However, understanding the process of making a novation can give individuals more control over their development, as well as a road map for increasing their contribution to the organization. When everyone in an organization understands the Four Stages$_{SM}$ Model, they will share a language for discussing development and growth, even when promotions are infrequent. The Stages Model makes clear the rules of a game that has always been played.

CHECKING FOR REALITY

While most organizations currently emphasize the importance of individual initiative in career development, no career exists in a vacuum. In the past, supervisors have been expected to provide coaching and feedback about a person's development goals; in fact, the supervisor may have been largely responsible for creating the development plans for her direct reports. However, downward feedback alone reinforces a hierarchical mindset that clashes with the cultural values most leaders are attempting to create. Moreover, as management spans of control broaden, the quality of downward feedback is jeopardized; it's difficult for one person to observe and relay detailed performance information about 20 or more people.

Feedback from multiple sources provides people with information about how well

---

[3]Unpublished research, Novations Group, Inc., Provo, Utah, 1998.

they are contributing to the achievement of team and organizational goals and how they can further increase their value to the company. It provides direction on what skills or competencies are most important. The individual absorbs that feedback and prioritizes his plan based on a combination of feedback from others and his own goals and priorities.

Many organizations now provide 360-degree or multi-source feedback reports, which are computer-generated, anonymous summaries of how a person rates on selected dimensions. When feedback is received anonymously from more than one source, it allows managers to approach employees as a coach providing assistance rather than as a critic imposing judgment. Feedback can help reveal blindspots as well as highlight strengths that may have been neglected.

Consistent feedback from a variety of sources is also more compelling and more difficult to rationalize. It's difficult to blame negative perceptions on the unfairness of one's boss when co-workers, direct reports, or customers corroborate the results.

INDIVIDUAL DEVELOPMENT IN ACTION: A CASE STUDY

One of the world's largest chemical companies applied all of these principles in a way that gave its technical and professional employees worldwide a way of planning and achieving their development goals.

In the early 1990s, the company evaluated its succession planning and development systems and discovered that moving people through a series of jobs didn't necessarily result in the development of the leadership skills they needed. At the same time, the company conducted a major study that concluded that the leadership qualities that had been rewarded in the past weren't the same as those that would be needed to make the company successful in the future. It also became clear that the company would be more successful if development opportunities were available to everyone in the technical and professional ranks, even if they didn't want to move into management or weren't identified as "high potentials" early in their careers.

The company wanted to provide clear expectations about what skills and abilities would be needed for the future, and also wanted to provide a process for employees to meet those expectations over the long term. As a first step, they identified in behavioral terms the leadership factors that had been identified as critical to the company's future. Each of these dimensions was described by each of the Four Stages$_{SM}$, so that high performance was described in a way that was achievable by anyone in the company, regardless of experience level. Describing the dimensions by stage also provided clarity about where development efforts should be focused. The diagram below shows the description of one of the leadership dimensions by the Four Stages$_{SM}$.

Customer Focus

Helping and Learning	Contributing Independently	Contributing Through Others	Leading Through Vision
Has a basic knowledge of customers (internal or external) and their needs; is responsive to customers	Actively seeks customer input; anticipates customers' needs and effectively meets them; seeks feedback to ensure customer expectations are met	Helps others understand customer needs; develops effective partnerships with customers; models good customer relations; seeks ways of improving service and building the customer base	Fosters culture and organization systems that entrench customer service as a key value

The company then created a development planning process whereby employees identified their own needs and priorities, their long-term goals in terms of contributing to the company, and potential derailers that might cause problems in their careers. From the list of 23 leadership dimensions, each employee selected three to highlight as strengths and three to focus on as development gaps. This selection was based on the employee's self-assessment and on an assessment from his or her manager. Although multi-rater feedback was not a requirement in all business units, a feedback process was readily available. Employees were encouraged to solicit a 360-degree feedback on a regular basis in order to measure development progress or monitor changing role demands.

Each employee received training on the Four Stages_SM Model, Derr's Career Orientations Model, and the development planning process. The action planning process emphasized identifying or creating stretch assignments in one's current role, rather than focusing on future jobs. The action plans were linked directly to the development gaps. Employees also received training on how to initiate and lead a development discussion, and have an opportunity to practice leading a discussion based on their plans with someone other than their supervisor.

This employee development system not only provided employees with tools for managing their own careers; it also changed the culture of the company to one where employees accepted and valued the opportunity to control their own destiny.

CONCLUSION

Self-directed career development is one of the new corporate realities. Taking responsibility for one's own career provides options and opportunities that probably wouldn't have been available before. However, without effective tools, most employees aren't equipped to plan and manage their own careers. A few simple concepts and tools can help employees grow in ways that will increase both their own satisfaction and the productivity of the organization.

ABOUT THE AUTHOR

Kathy Buckner is a senior consultant specializing in competency-based HR systems. Since she joined Novations, she has been involved in the development and roll-out of employee development and performance management systems at DuPont, Champion International and MetLife.

Before joining Novations, she consulted with firms of all sizes on a variety of issues. She was the director of the Utah Small Business Development Center and was an instructor at Brigham Young University's Marriott School of Management. She has published a number of books and articles on employee development and other management issues.

Ms. Buckner holds a master's degree in organizational behavior.

ABOUT THE PRESENTER

Kristen Knight is a consultant with Novations Group, Inc., specializing in the development of multimedia training programs. For the past five years, she has worked in managerial, training, and consulting positions with companies such as Nordstrom, Lightspeed Computer Systems, and Bell & Howell Powersports.

Kristen received her master's in organizational behavior from Brigham Young University.

Effective Communication Skills for Engineers

S. CERRI

ABSTRACT

If technical professionals really know how to communicate, they can accomplish any job more effectively. Effective communication is the key to successful outcomes. Technical professionals must be trained in the discipline of effective communication just as they are trained in the discipline of electrical and electronics engineering. Fortunately, our understanding of the process of human communication has finally achieved a level of structure that allows training. This paper presents a step-by-step approach to the discipline of effective human communication.

Over the past 20 years, research has provided insight into the processes of human perception and communication. These insights have led to the development of tools and procedures that increase the effectiveness of that communication significantly. The communication process is divided into seven distinct steps. When taken in order, these steps give the speaker the ability to communicate and influence a wide variety of people. Engineers and technical professionals in the IEEE environment can use the "7-Step Effective Communication Process" to increase their communication effectiveness. By doing so they will stand out as being more capable and more competent at influencing, leading, and conveying ideas and concepts to their colleagues and to others.

The 7-Step Effective Communication Process allows the speaker to understand the structure of the communication that would be most beneficial for the listener. Once this has been established, the speaker then builds rapport and begins the process of understanding the listener's paradigms. Once the listener's paradigms are understood, the speaker can then send the desired message and then determine whether the message was received as intended. This process is fast and can be done in the course of casual conversation, in the midst of an important meeting, or from a platform presentation. It can be performed in a large group or one-on-one. It is easy, quick, efficient, and effective.

Steven Cerri, Director of Corporate Training, Eltron International, Inc.

INTRODUCTION

Our technical world is becoming more complex while our need to communicate effectively with a wider variety of people is increasing. The next millennium requires technical professionals who can communicate effectively and skillfully in a variety of situations and with people with diverse cultural and professional backgrounds and abilities. This is our present requirement and our future need.

However, the typical engineering graduate has spent four to five years studying physics, dynamics, chemistry, electronics, and engineering in an attempt to prepare for a productive career in the field of engineering and often not even one class has been devoted to human communication processes. The engineering disciplines require a certain mode of thinking. The analytical processes required to succeed as an engineer do not automatically lend themselves to effective human communication. Very few engineering students have the insight to enroll in communication classes. Very few engineers even believe that communication is an issue for them until they are in the work environment and are faced with what seems to be an inability to connect with and influence people. For the typical engineering student, effective communication skills are assumed to come along as a process of human maturation.

What would be the benefit if communication tools and techniques were taught to engineers? Imagine engineering graduates who could effectively convey their ideas and concepts and values to a wide variety of colleagues and customers. This would be a boon to the engineering and business communities. Specific communication tools and techniques exist to help technical professionals achieve this outcome. When these techniques are applied, effective communication is all but guaranteed.

WHAT IS COMMUNICATION EXCELLENCE?

Communication excellence is based on the natural structure of language and how the human brain processes verbal and non-verbal information. When we have a clear and sound understanding of the human processes of perception, communication, and cognition, then the process of communication becomes a predictable closed system with input, output, and feedback. When the communication process is understood within the framework of a "system," the process lends itself to analysis and understanding rather than being a "hit-and-miss" process.

Approximately 20 years ago, research began that asked the questions, "What is effective and excellent communication? Can this excellence be modeled? If it can be modeled, can its components be determined and taught so that other people can become excellent communicators as well?" The answer to each of these questions was an emphatic "yes," and the answers gave birth to a discipline that now provides the tools that allow anyone to be an excellent communicator. These tools have been assembled into the 7-Step Effective Communication Process.

HUMANS AS SATELLITES

Effective communication is the process of understanding that each human being is like a satellite moving through the universe, gathering data about that universe—the universe of life. Just as a satellite uses its instruments to gather data, people use their senses to gather data about the world around them. The data they collect from their senses are then integrated into a meaningful "map" of reality that they then use to make choices and move through the world. This "map" of reality is a filter that is used to focus attention on and filter information from the world around them. This "map of reality" is used to filter all other data as it enters through the senses, including communication from other people. If the filters are fully "operating," then the communication process can be difficult. The listener will "filter" the incoming messages based on the "biases" of those filters. If, on the other hand, the filters are not engaged, the messages coming in to the listener will be absorbed without the prejudice of those filters.

Therefore, the definition of effective and excellent communication is the ability to communicate so that the listener's "filters" are not engaged or at the very least, minimally engaged. If the listener's "filters" are not engaged during the communication exchange, the speaker can have a high probability that the listener will actually "hear" the communication in an unbiased fashion.

WHO IS RESPONSIBLE FOR EFFECTIVE COMMUNICATION?

Some people would say that the sender (speaker) is responsible for effective communication while some would say the listener is responsible. Yet others might say that both parties are actually responsible. The author believes the true party responsible for effective communication is the sender, because only the sender knows exactly what the communication is meant to convey.

HOW DO ENGINEERS AND MANAGERS COMMUNICATE DIFFERENTLY?

During college, engineering students are taught that the context of a communication is less important than content. The goal is to arrive at the correct answer; knowing the answer to an engineering problem is what is demanded.

In the business world the answers are multiple and less a function of correctness and more a function of influence and power. Often a "less right" answer will hold sway over "the right" answer if the proponent has more power or influence or if the proponent can convey the message more persuasively. This does not mean we should condone such a process; it is merely a fact of life.

In the business world, context often is as important or even more important than content. Therefore, the technical professional who wants to be a successful communicator must understand the role of context as well as content. A guiding principle for effective communication is, "never communicate content without first establishing the appropriate context."

COMMUNICATION PITFALLS: DELETION, DISTORTION, AND GENERALIZATION

During any communication, speakers and listeners perform mental and verbal processes that either enhance or obstruct the effectiveness of the communication. These mental and verbal processes are deletion, distortion, and generalization. Deletion is the elimination of useful or important information. Distortion is the process of changing the meaning of a communication. Generalization is the expansion of a communication message to include other, unrelated applications. These three communication "violations" impede effective communication because they change the resulting communication into something different than originally intended. By understanding the roles of deletion, distortion, and generalization, communicators can remove the negative effects of these three processes.

COMMUNICATION "REPRESENTATIONAL SYSTEMS"

People "represent" their experiences in their minds by using the five senses. Each sense allows the human brain to collect data about the world. On the conscious and subconscious levels, the human brain "represents" data in a way that is tied directly to the representational system through which the information was brought to the brain. That is, visual information is stored as pictures. Auditory information is stored as sounds and the sounds of words, while kinesthetic information is stored in the nervous system as feelings and emotions. In this way, much of the sensory data that people accumulate over time are actually maintained in the forms in which they were originally acquired. Although other transformations can take place over time, the key point is to understand that the human mind actually stores information in representational systems correlated to the five senses.

THE THREE WAYS WE COMMUNICATE 100 PERCENT OF THE TIME

For people in the business world, most communication takes place using the following three representational systems: visual (sight, pictures); auditory (sound and words); and kinesthetic (feelings, sensations). Olfactory (smell) and gustatory (taste) systems do not enter into the business communication process often. Although the five representational systems obviously make up 100 percent of our communication environment, visual, auditory, and kinesthetic data usually make up the majority of our daily professional communication.

BUILDING RAPPORT—HOW FRIENDS COMMUNICATE

Just as two transmitters transmitting at the same frequency can increase their combined amplitude, two people communicating in the same representational system can communicate very well. Conversely, two transmitters transmitting 180 degrees out of phase will subtract from each other's amplitude. In the same way, two people communicating in different representational systems will find communication to be difficult. Friends are actually friends, to a large degree, *because* they communicate in the same representational systems most of the time.

THE "7-STEP COMMUNICATION PROCESS"

The "7-Step Communication Process" defines a strategy for effective communication that can be used in any situation. The steps are presented here.

1. *The first step* in effective communication is to understand which representational system the listener is operating within. By understanding the real-time representational system, the speaker can "match" that system, thereby ensuring that the communication will be in an acceptable mode to the listener. This specific technique of using verbal and non-verbal communication can effectively remove most if not all the filters from the listener's side of the conversation. This allows the communicator to be much more effective.

2. *The second step* in effective communication utilizes the techniques of mirroring, matching, pacing and leading of verbal and non-verbal communication cues. These tools allow the speaker to build unconscious rapport with the listener, thus placing the listener in a more receptive mode. This process reduces the effects of the listener's "filters" and allows the speaker's messages to be received more openly.

3. *The third step* in effective communication is to uncover the listener's complex maps or paradigms of reality. This questioning process can be quick and elegant and ensures that the speaker understands the paradigms that must be matched in order to initiate and continue effective communication.

4. *The fourth step* in effective communication is for the speaker to send the message. At this point, rapport has been established and the communication process can proceed in such a way that the sender can present ideas and issues and the listener will be receptive to them. This does not necessarily mean that the listener will agree with the sender's position, but at the very least, that the listener will hear the sender's message in an unbiased way.

5. *The fifth step* in effective communication is for the sender to "check" to determine whether the message was received by the listener as intended. In fact, it is important to point out again that the responsibility for effective communication rests with the sender; only the sender knows what the intended message is. Depending upon the filters engaged by the listener, the message could be interpreted in a variety of ways.

Therefore, the fifth step is for the speaker to ask questions and watch for non-verbal cues to determine whether the message was received by the listener as intended.

6. *The sixth step* in effective communication is to go back to steps 1, 2, or 3 to set the stage to send the message again if it was not received as intended.

7. *The seventh step* in effective communication is to send the next message using steps 1, 2, or 3 if the message was received as intended.

The seven steps form a closed, feedback-look system with input, output, and feedback. With these fundamental tools for effective communication, the sender can communicate ideas and concepts more effectively and can influence people more elegantly. Within the IEEE environment, the typical technical professional will be capable of speaking and communicating more effectively with other technical professionals, as well as with managers, accountants, lawyers, contract personnel, and customers.

SUMMARY

Excellent communication is a process and an ability that can be learned. Steps and tools exist for communicating more effectively with a wide variety of people. These tools are applied using the "7-Step Effective Communication Process." Each step in this process leads the speaker along a path that increases the ability to convey information to the listener.

"The 7-Step Effective Communication Process" is summarized as follows:

1. Determine the representation system strategy of the listener.

2. Build rapport with the listener.

3. Step into the listener's representational framework and ask questions to determine the listener's structure of reality.

4. Send your message in conformance with the listener's map of reality and representational system.

5. Check to determine whether the listener received the message in the way in which it was intended.

6. If "NO"—the message was not received as intended. Go back to Steps 1, 2, or 3 and send the message again (Step 4).

7. If "YES"—the message was received as intended. Go on and send the next message.

By using these seven steps and the processes imbedded in them, technical professionals can communicate much more effectively.

REFERENCES

Andreas, Steve and Faulkner, Charles. 1994. NLP The New Technology of Achievement. New York. William Morrow and Company, Inc.

Bandler, Richard and Grinder, John. 1975. The Structure of Magic. Palo Alto, California. Science and Behavior Books, Inc.

Bandler, Richard and Grinder, John. 1979. Frogs Into Princes. Moab, Utah. Real People Press.

LaBorde, Genie. 1983. Influencing With Integrity. Palo Alto, California. Syntony Publishing.

O'Connor, Joseph and Seymour, John. 1990. Introducing Neuro-Linguistic Programming. Hammersmith, London. Mandala Publishers.

Richardson, Jerry. 1987. The Magic of Rapport. Cupertino, California. Meta Publications.

Yeager, Joseph. 1985. Thinking About Thinking With NLP. Cupertino, California. Meta Publications.

ABOUT THE AUTHOR

Steven Cerri successfully made the transitions from entry-level aeronautical engineer to program manager, to director, to vice president, to general manager, to entrepreneur, to corporate trainer. He has been a manager, leader, entrepreneur, and trainer for the past 20 years, and has successfully "transitioned" many top-level managers and directors; many have managed multi-million dollar programs and have gone on to build their own successful technical businesses.

Mr. Cerri currently appears on television stations around the United States (including KCET) twice a week to discuss management in the 21st century. He has a bachelor's degree in aeronautical engineering, a master's degree in geophysics, and a master's in business administration. In addition, he is a master practitioner of neuro-linguistic programming.

For more information about communication excellence and the 7-Step Effective Communication Process, contact Steven Cerri at (805) 578-7124.

Transitioning from Technical Professional to Manager

S. CERRI

ABSTRACT

Leadership is often defined as learning how to get people to do the things you want them to do. The real question, however, is how do we motivate people? In the next millennium, our leaders will have to motivate a more highly educated, more independent, and more diverse group of employees than ever before. Where are we to get our new technical managers?

Many successful engineers are promoted to various management positions because they have been successful in their previous technical positions, not because it is known that they will be good managers. Promotion to management is often a reward for a technical job well done. However, the traits and abilities that make a technical person successful in a technical position are most likely not going to be the traits and abilities that will make them successful as a manager. What are the traits of a successful manager and what is a technical professional to do to make the successful transition from the technical world to the management world?

The paradigms of a good technical professional and the paradigms of a good manager are juxtaposed. The values and beliefs about what is important for these two professions are basically at opposite ends of the behavioral spectrum. In order for technical professionals to make the transition to manager, they must modify years of college preparation for the sciences and align their attention to the human side of the business equation. While most technical professionals receive satisfaction from their own accomplishments and their contributions to the technical team, the successful manager must receive satisfaction from the successes of those he or she leads. Without this shift in values, the technical professional can only advance a short distance up the management ladder.

This paper outlines some of the shifts that must be made in order to make a successful transition from technical professional to a successful manager.

Steven Cerri, Director of Corporate Training, Eltron International, Inc.

WHAT IS A TECHNICAL PROFESSIONAL?

The typical engineering graduate has spent four to five years studying physics, dynamics, chemistry, electronics, and engineering in an attempt to prepare for a productive career in the field of engineering. As the young engineer achieves success in the technical world, the typical reward for such performance is a promotion to manager. The tacit assumption underlying this promotion is that if the person is successful as an engineer he or she can certainly succeed as a manager.

In college the typical engineering student spends no time at all studying human communication principles and management techniques. The expectation is that those abilities are learned by everyone as they mature and grow and this capability need not be developed through formal education. Technical graduates have spent years in college learning how to analyze technical problems. They have dealt with concepts and theories. Attention to human communication and the issues surrounding the management of people has, for the most part, been non-existent.

WHAT IS IMPORTANT?

Most technical professionals feel they have not completed a good day's work if they have not produced a graph, completed an analysis, written lines of code, designed a circuit board, or done some sort of analytical process. Good managers, by contrast, may never write a line of code and may never perform an analysis during a business day. Instead, they may spend most of their day talking to people, attending program meetings, dealing with personnel conflicts, interviewing prospective employees, completing performance reviews, determining the future direction of their organizations, dealing with operations issues, coordinating technical resources, listening to insurance presentations, predicting the required size of a new building, and deciding how work flow is to be divided.

The average engineer, performing the tasks of the average manager, would end the day complaining that he or she had accomplished nothing of importance. He or she might conclude that, "I just spent the whole day talking to people." Making a successful transition from technical professional to manager requires that the technical person actually change his or her "mental maps" of what is important and significant. Without this change in focus and in values and beliefs, the transition from technical person to manager is impossible.

WHAT DOES IT TAKE TO MOTIVATE PEOPLE?

Many different styles can be used to motivate people, including fear, intimidation, authority, incentives, empowerment, and participatory management. However, the question is no longer "can we motivate people," but rather, how do we motivate people so that we can get their best efforts, their creativity, and their talents? The answer to this question does not lie in negative strategies. Such strategies only give the organization the robotic actions of the employees. The creativity and best efforts of each person on the team are only provided when leaders inspire and move people to belong to something bigger than themselves. Without an understanding of human behavior and the processes that lead to establishing aligned values and beliefs, managers are unable to manage and lead effectively. One of the aspects of human motivation is known as "sorting."

A QUESTION OF SORTING

As people move through the world, they determine what information to focus on and what information to throw away. This process is called "sorting." People move through the world and "sort" the data and information they receive through their mental filters such that they keep some data and they ignore some data.

Generally in the professional world, we can place the data sorted into five different categories: people, places, things, knowledge, and activities. That is, a person who sorts with a priority for *people* will not be as concerned about what they are doing (i.e., activities) but will be most concerned about those with whom they are doing it (i.e., people). Conversely, a person who sorts first for *knowledge* will not be as concerned about where they are (i.e., places) or who they are with (i.e., people) as long as they are gaining *knowledge*. People who live in beautiful areas or locate to specific areas of the world (i.e., places) and find whatever job they can get are sorting by *place*. Where they live (i.e., places) is much more important than what they are doing (i.e., activities).

By this analysis, it is easy to see that, as a general statement, most technical professionals sort by *knowledge* and *activities* first. They are most concerned about what they are doing and what they are learning. People, places, and things are going to be sorted for on a lower priority.

We can likewise make a general statement about managers. That is, successful managers often sort with *people* and *activities* as the top two sorting preferences. The effective manager focuses on developing people because their success is dependent upon the accomplishments of others. Also, the successful manager is required to perform certain necessary *activities*, such as meetings, which may not be challenging or may not contribute to significant learning. Places, things, and knowledge may be in varying priorities below the top two: people and activities.

Obviously, these statements are generalizations, and we can always find exceptions to them. The point however, is that the average technical professional is focusing (sorting) for very different data in the world than the professional manager. In almost all cases these sorting processes take place on the subconscious level. Without raising the sorting process to a conscious level, change is difficult. The conclusion is that the technical professional must change the sorting priorities in order to be a successful manager.

Sorting is only one aspect of human perception and cognition that is different for the technical professional and the successful manager. A variety of human "perceptual processes" separate the successful technical professional from the successful manager. Previously, the change process was left to time, experience, and often, to chance. Understanding these processes allows people to transition to management much faster and with greater success.

SUMMARY

The transition from technical professional to successful manager is not simply a matter of dictating a change in behavior. Behavior or action is the result of a perception about the world. If we want to change behavior we must first influence perception. If we want technical professionals to behave like managers, we must first change the way they "think" about the world. We must influence their values and beliefs and refocus their attention. To require that the technical person "just change" is naive. Several aspects of human perception and cognition must change in order to achieve a successful transition to management; one of those aspects is known as "sorting." When they change their sorting priorities, technical professionals can make the transition to successful managers more easily.

REFERENCES

Andreas, Steve and Faulkner, Charles. 1994. NLP The New Technology of Achievement. New York. William Morrow and Company, Inc.

Bandler, Richard and Grinder, John. 1975. The Structure of Magic. Palo Alto, California. Science and Behavior Books, Inc.

LaBorde, Genie. 1983. Influencing With Integrity. Palo Alto, California. Syntony Publishing.

ABOUT THE AUTHOR

Steven Cerri successfully made the transitions from entry-level aeronautical engineer to program manager, to director, to vice president, to general manager, to entrepreneur, to corporate trainer. He has been a manager, leader, entrepreneur, and trainer for the past 20 years, and has successfully "transitioned" many top-level managers and directors; many have managed multi-million dollar programs and have gone on to build their own successful technical businesses.

Mr. Cerri currently appears on television stations around the United States (including KCET) twice a week to discuss management in the 21st century. He has a bachelor's degree in aeronautical engineering, a master's degree in geophysics, and a master's in business administration. In addition, he is a master practitioner of neuro-linguistic programming.

For more information about transitioning from technical professional to manager, contact Steven Cerri at (805) 578-7124.

Successfully Speaking: Winning Government Orals by Giving Memorable Speeches

R. L. CRANSTON

INTRODUCTION

This presentation focuses on the difficulties of giving a government briefing when time is short and the pressure is intense. It will help attendees prepare for that day when the boss calls and says, "John or Jane, next week you are going to Washington to give a briefing on why we can do the best job fielding the new XXX System." It will provide insight into how doing business with the government has changed.

In the past, briefings were made by people who were good speakers, such as corporate managers and consultants. In today's environment, the government requires companies to have briefings presented by the managers and engineers who are proposed to manage or perform the work. This means that the proposed program and project managers must look the selection committee in the eye and convince them that they represent the best buy for the buck. The engineering leads must convince the selection committee that they are smart and capable people. Briefing scenarios often include a technical problem that the team has 15 minutes to review and solve, and then make an immediate oral response. In addition, that briefing team must be prepared to respond to questions.

This presentation will draw heavily upon the experiences of Systems Resources Corporation (SRC), a provider of Federal Aviation Administration (FAA) and Department of Defense (DOD) services. SRC's presentation experiences have resulted from its ability to write successful Screening Information Request (SIR) responses; companies won't make it to the briefings phase unless they can prepare and submit top-notch proposals and reports. In addition, this presentation will describe many of the successful speaking techniques expounded by Toastmasters International, an organization whose mission is to develop speakers who can be successful in any given situation.

ENGINEERS MUST BUILD PUBLIC SPEAKING SKILLS TO SURVIVE

Corporations and small companies are now entering an environment that will re-

Robert L. Cranston, Systems Resources Corporation

quire their philosophies on winning government business to change. Not long ago, engineers did their engineering and made presentations to bosses, but did not get overly concerned about giving quality briefings. After all, quality was in the value of their thoughts or designs. In this new environment, engineers will be preparing and presenting proposal responses more often to a government selection committee; their presentation skills will be compared to those of other bright and motivated engineers. The days of "um-ing" and "ah-ing" through a speech and still being successful are ending quickly. Shuffling through a pile of slides and reading them to the audience will soon force your company out of business. Ultimately, the people who have the training to present convincing arguments will earn the business. This briefing describes the nature of the new contracting environment and provides fundamentals of speaking. These fundamentals will help us enjoy the power of speech, improve our self-esteem, open up leadership positions, and increase our value within the marketplace.

Successful presentations frequently depend on how well we analyze our potential audience. I once prepared a speech for a non-profit organization in the Atlantic City area. Based on discussions with the coordinator, I anticipated the evening's audience to be a group of young, highly motivated businessmen. Upon arriving, I found that most of the audience consisted of retired businessmen who were more interested in golf and sports than on getting ahead or preparing for a successful career. During supper, I mentally revised my presentation, making it appropriate to the evening's audience. As a result, the speech was highly successful.

The more you can learn about your audience, the better you can pitch your argument. During one of SRC's presentations, my team learned that the customer was more interested in relationships than with our technical capabilities. Consequently, we tailored our presentation and materials to achieve this goal. We dressed down, we limited our presentation to leafing through paper slides, we listened a lot and we asked a lot of questions. We guessed right and our people came across as being a team—likable, perceptive, and capable of listening.

Government audiences will vary from being polite and professional to being downright hostile, especially when they favor the company currently performing the contract. When preparing your presentations, be prepared for indifference *and* for highly challenging or emotional questions. When observing indifference, speakers must be able to turn the indifference into interest. When presenting results of a technical problem, the briefing team needs to consider the interests and technical levels of the audience. The panel selection engineer might have a circuit board mentality while the contract officer's technical representative might have a system-level understanding. Your team's oral presentation must demonstrate your team's ability to think quickly and to give a well-organized presentation that uses an appropriate level of technical language.

Presenters must consider what image will impress the government selection committee. To be successful, briefers need to project confidence through dress, voice and attitude. Traits that can influence an audience include intelligence, ability to think clearly, tact, enthusiasm, subject matter knowledge, and yes, knowledge of effective speaking. To maintain a professional image, do not take things personally; for example, do not respond to slightly condescending remarks that may or may not have been intentional. Another word of caution: never become apologetic or defensive if you make a delivery mistake. Unless it is brought to the audience's attention, your nervousness might go unnoticed and a typing error or slightly smudged slide may be overlooked. Speakers must remember that once they are in front of the government selection team they are selling their company as well as themselves. Remember, therefore, that image is important.

Projecting confidence requires one to control nervousness. Controlling nervousness requires a presenter to practice, practice and practice. Nervousness and bad speech mannerisms will kill an otherwise fine presentation. Companies need to both budget and plan for preparing presentations. Speakers need to practice vocal variety to show enthusiasm, energy, and confidence. Speakers need to eliminate their "ums" and "ahs," since these are characteristics of untrained speakers. If a pause is required, take a natural pause to collect your thoughts. Speakers often use pauses effectively to enhance or make a statement.

A well-organized, thought out briefing is necessary to minimizing questions and to selling your team's strengths. To be competitive, all briefing items included in the SIR need to be addressed. One of the hot questions I addressed recently was the requirement for our company to be ISO-9000 certified. I was able to persuade the audience that our quality program would exceed their expectations for this effort.

A quality briefing begins with clearly defined objectives that outline your company's philosophy. Philosophies are important because they indicate how your company will likely approach problems and will manage employees and quality. For example, if the company wants to emphasize value over cost, your presentation will need to detail how your company offers value and how your customers benefit. For example, you might contrast your company to a cheaper provider, citing high personnel turnover or the failure to attract people because wages are too low or benefits are too meager as weaknesses. Your company's low turnover, quality products, and greater productivity can be a strong argument as to why the customer should avoid a low-cost provider. Naturally, this implies that your team knows the competition. Similarly, briefers can make arguments for quality management and adequate staffing. Many a skillful argument has changed history, kept a man out of jail or kept a politician in office.

Government oral presentations often involve several speakers. Briefers need to organize their presentations to ensure proper introductions of all personnel. The briefing needs to include the typical speech structure of having an opening, body, and conclusion. Within this format, the presentation should consist of ideas that flow from the introduction into the summary. When preparing effective briefings, plan to tell the audience what you are going to tell them, tell them, and wrap it up by summarizing what you have told them. The first 30 words catch the audience's attention and make them want to listen; the last 30 words contain the most important message you want them to remember.

The SIR will normally contain questions that you will need to address during the presentation. Answers to the questions need to support the contention that you and your company can do the best job and in the process offer the best value. Integrate into your material the philosophies that you want the audience to remember. Consider carefully the idea of trying to impress an audience with how much money your company makes before you include such information in a briefing. The customer may think, "Since they are so successful or big, will they consider me important?" Your briefing should contain information sufficient to convince the potential customer that your company has a solid reputation for providing quality products and services to a number of customers.

Usually, the briefing team will have one hour to make a presentation. Therefore, convincing the audience that you understand the requirements and can perform successfully is a major objective. When preparing a briefing, consider:

- Should the technical information be summary or in-depth?
- Will there be opposition to certain ideas?
- Is the information relevant to the briefing objectives?
- Will any of the information likely lead to unfavorable questions?
- Is the minimum amount of information presented to convey the main ideas?

The conclusion needs to capsulize the major points of the presentation. A skillful summation will leave in the minds of the audience the reasons why your team should be selected for this particular job or task.

Following the presentation, the government selection committee will have a question-and-answer period. The briefing team will need to be prepared to respond to any material contained in the SIR, information brought out during the briefing, or information related to their company's previous performance. The team will need to project the confidence that they are quite capable and motivated to hit the street running. Break complex questions down and watch out for bait-type questions that are designed to mislead or even embarrass. To present your briefing and respond to questions successfully, ensure that all team members understand the audience, speak to objectives, practice in realistic settings, and understand the destructive effects stress can have on speakers.

PLANNING AND PRACTICE ARE KEY

Successful companies must have planning and budgeting strategies for developing engineers' speaking skills. Like most people, engineers would prefer to eat nails, wrestle snakes, or face lions rather then give presentations. In front of an audience, the adrenaline races, the cortisol rises, and rational thinking diminishes. When cortisol rises, an inexperienced speaker will forget, misrepresent, or articulate information poorly. To be competitive, companies need to encourage their engineers to become involved in professional societies where they can exercise speaking and leadership skills. Poor strategic planning can result in having too small a speaking pool or unhappy and burned out presenters. To minimize stress and promote speaker morale, companies need to frame the speaking environment positively and compensate appropriately. Otherwise, the company will not find takers for speaking assignments. Subsequently, management will be most unhappy.

Speaker evaluations are essential to preparing for successful briefings. However, when critiquing speakers, corporate personnel need to provide constructive criticism in a positive manner. Occasionally, evaluators feel they are doing a great job if they can point out every little fault in a speaker's argument or delivery. When this occurs, the speaker's attitude can quickly turn rebellious and he or she might resent having to present now or ever again. This is especially true if the briefer spent the previous night researching material. Before making presentations, speakers need to feel confident and refreshed, not fatigued or "beat up." Neither corporate management nor individual speakers will be satisfied with the returns gained by stressed out presenters.

The government often requires copies of presentations several days before the big event. In addition, contractors are advised of the availability of projectors at the briefing site. During one presentation, the only materials allowed into the briefing room were the people doing the potential work, copies of briefings, and clear transparencies. The team used the clear transparencies to prepare a speaking outline for the technical problem.

Briefers often make two bad assumptions. First, they think they can read from their slides. When briefers read from slides, they draw a curtain between the audience and themselves. Further, most audiences can read. A second frequent problem occurs when the briefer's presentation gets ahead of or behind the information shown on the slides. During slide presentations, briefers must maintain eye contact with the audience, think constantly, and be able to read body language relative to the presentation so they can make adjustments as necessary.

It is helpful to use video cameras during briefing practice sessions, since they will get the presenter accustomed to being videotaped. The purpose of using video equipment is only to get speakers comfortable with recording equipment, not to embarrass them into doing a better job. And as the presentation date approaches, practice sessions should become as realistic as possible. Create a fearful environment and presenters will perform poorly. Create a positive environment and their performance will soar.

CONCLUSION

The new government acquisition strategy has created a wealth of opportunities for engineers, as it ties credentials to positions. In the future, companies will have fewer opportunities to skirt credential requirements because government agencies are attempting to determine up front who will be managing and leading their work. In the meantime, it behooves engineers to prepare for the challenge of presenting effective briefings. Engineers have built highways, computers, communications systems, and air traffic control systems. Now engineers will need to formulate and present effective arguments for why they should be the next government provider. Prepare for tomorrow by developing strong speaking skills today. When preparing for your next presentation, do thorough research, organize carefully, and practice. In order to be successful, people need to confront their speaking fears and develop good speaking habits. In the process, they will become winners.

Coaching for Contribution: The Leadership Behaviors That Make a Difference

R. CUTADEAN

ABSTRACT

For most people, development in large organizations has always represented a challenge. In today's business environment, it has become even more difficult. Ongoing downsizing, restructuring and re-engineering have created an environment of "do more with less." If you are not surrounded by people who are creative, motivated, skilled, and eager to take on new and more challenging roles, your organization will have difficulties meeting the demands (competition and change) of the information age.

Unfortunately, too many people view *developing others* and *getting results* as mutually exclusive or on opposite ends of the same continuum. On the contrary, in today's information-driven economy, coaching is the most powerful vehicle available for achieving results and building organizational capability. Coaching helps you increase the level of contribution within the organization by achieving results through others, which at the same time helps individuals grow and develop.

Over five years of research by Novations has resulted in the identification of six skill clusters (and the corresponding 45 behavioral indicators) that significantly impact an individual's ability to be an effective business or corporate coach. This paper will focus on summarizing the research findings and identifying the coaching behaviors that make a difference.

INTRODUCTION

Today's organizations expect you to be committed to the concept of life-long learning. Unless other people see you as committed to your own learning, it will be hard for them to accept you as a coach. If you want direct reports and peers to be open to coaching, you will need to model the learning process.

Coaching is not a minor responsibility for leaders to fulfill when the pressures for re-

Ron Cutadean, Novations Group, Inc.

sults subside. Coaching is also not optional. We are all coaches, at least in a passive sense, whether we like it or not. The only choice we have is whether to be a good coach or a bad coach. The right choice first requires you to have an increased awareness of the key competencies associated with high-performing coaches. This paper should accomplish that. The second requirement is that you must become deliberate in the application of these competencies to your work environment. That is when the real fun and work begins!

COACHING DEFINED

Coaching is essential to the process of developing people. On-going coaching, however, is dependent on the existence of having both willing and skilled coaches as well as open and active learners who want to be coached. Coaching is therefore defined as: *The collaborative process of helping others develop specific competencies that will contribute to their achieving superior individual and organizational performance.*

COLLABORATION BUILDS "WIN-WIN" OPPORTUNITIES

High-performing coaches recognize that over the long-term, both individual and organizational needs must be met. Coaching is not a zero-sum game. If the individual's needs are not met, his/her willingness and ability to meet the goals of the organization will shrink. On the other hand, if the organization's goals are not met, its ability to provide on-going developmental opportunities for the individual will shrink.

High-performing coaches help others get on **TOP**[SM] of their game, in an effort to have them achieve more frequent or sustained "career bests." A career best is a sus-

FIGURE 1. Coaches help others get on TOP[SM] of their game.

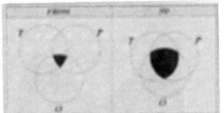

FIGURE 2. Coaches help others maximize the intersection.

tained period of time when an individual is maximizing their personal/professional satisfaction in connection with fulfilling their job responsibilities for an organization.

The **TOP**^SM Model depicted below graphically illustrates this important relationship between employees and the organization. Successful coaches collaborate with others in an attempt to realistically and positively impact the relationship between an individual's talents (**T**) and passion (**P**) and how that can best be played out in their current (as well as future) positions in the organization (**O**).

In order to optimize the intersection of the three circles, the coach first needs to communicate and provide a contextual perspective of the organization's business needs (Organization), support systems, and the individual/team performance objectives to those he or she coaches.

In addition, the coach, as well as the individual (learner) receiving the coaching, must be extremely clear about what motivates/interests (Passion) him or her and how this individual's competency strengths and weaknesses (Talents) impact his or her ability to achieve the organization's business objectives.

The biggest mistake coaches make is to assume that they know what an individual needs. These assumptions are often wrong. The only way to understand what others need is to promote open dialogue and by asking them. Optimizing the intersection of organizational and individual needs cannot be achieved unilaterally. It has to be collaborative. Both the coach and the individual must work toward the same ends.

Individuals will always have some degree of intersection between their Talents, Passions and the Organization's needs—even if they receive no coaching. High-performing coaches, however, significantly and deliberately employ key behaviors to maximize the intersection of these three important **TOP**^SM dimensions. The following diagram illustrates the increase in this so-called "sweet spot" or intersection, which visually depicts the role of today's business coach.

A FRAMEWORK FOR COACHING

The following diagram is a visual summary of the six competencies clusters required for today's business coach. The two outer rings build the foundation that fuels and magnifies the power behind the key behaviors and skills within the other four skill clusters.

FIGURE 3. A framework for coaching.

Ignoring these building blocks will make a coach work harder and less efficiently in these other four skill clusters, thereby achieving sub-optimal results.

ESTABLISH A CLIMATE FOR LEARNING AND TRUST

Coaching works best in an environment that encourages and values learning. Individuals are more apt to want to learn when they know their requests for help (or the coach's offer to help) will not be interpreted as a lack of confidence and/or an absence of potential. Because learning something new brings with it the risk of failure, successful coaches help build the confidence of others so that they take reasonable risks in the form of stretch assignments, new roles, and new behaviors.

Of the six, this is probably the competency cluster most taken for granted, probably because our research base shows that most people receive reasonably strong scores on the profile items. Unfortunately, if you are not vigilant in protecting the behaviors that impact others' perception of your trustworthiness as a coach, then the time it takes to rebuild your credibility is several magnitudes greater than the time it took to lose it!

The Behaviors That Make a Difference

High-performing coaches:
- ☐ Minimize differences in power or status between oneself and others
- ☐ Don't view requests for help as a sign of weakness

- ☐ Back others up when they make "honest" (developmental) mistakes
- ☐ Earn trust by not taking advantage of others
- ☐ Continually look for opportunities to be helpful to others

CREATE A CONTEXT FOR PERFORMANCE AND DEVELOPMENT

Effective coaches recognize and help others recognize that performance (meeting and exceeding job requirements) is inextricably linked to development (increasing the depth and/or breadth of one's competencies). Helping others see that there is a multitude of opportunities to develop and grow in their current assignment, in order to maximize the **TOP** ^{SM} intersection, is the primary challenge facing the business coach.

As a result of the increased emphasis on teams and teamwork, successful coaches recognize the importance of *Utilizing the Strengths and Diversity of Others*, a subset of this competency cluster. Effective coaches demonstrate respect for the ideas of others and expect others to do the same. Helping individuals to understand the challenge and importance of merging the needs of the business with the interests and talents of a diverse work force is the essence of the relationship between performance and development.

The Behaviors That Make a Difference

High performing coaches:
- ☐ Help others see how their career goals can be met in the context of their current job
- ☐ Consistently show an interest in the careers and development of others
- ☐ Provide frequent feedback to others
- ☐ Help others find challenging tasks to stretch their learning
- ☐ Help others find developmental opportunities that have a visible business impact
- ☐ Help others find roles/assignments that are consistent with their talents
- ☐ Help others discover or better understand their unique talents and interests
- ☐ Encourage others to stay open and learn from the diversity of ideas and perspectives
- ☐ Demonstrate respect for the ideas of others, even in disagreement
- ☐ Recognize that different things motivate different people

PROMOTE DISCOVERY AND SELF-RELIANCE

To increase the self-reliance of individuals, effective coaches must help them "discover" answers (how to develop key competencies) for themselves rather than telling or giving them the answer. The quote, "Give a man a fish and you feed him for a day; teach a man to fish and you feed him for a lifetime," captures the theme of this skill cluster.

Successful coaches are also adept at helping others *Identify the Development Opportunity*, which is a subset category within this broader competency cluster. A clear indicator that self-reliance has been achieved is when an individual can identify developmental opportunities without the assistance of a coach. Giving and receiving feedback are important behaviors for uncovering those opportunities for development and speeding the process to self-reliance. This does not mean, however, that the coach is not responsible for uncovering developmental opportunities as well. Listening and watching for developmental opportunities for others is an important way to leverage a coach's time and significantly increase personal/team contributions.

Our research also clearly indicates that high-performing coaches' perceptions of their skills and behaviors are closely aligned with the perception of others. This alignment is achieved by the coach continually seeking feedback from others on how his or her behaviors described here (as well as other behaviors) are personally impacting them in the accomplishment of work.

The Behaviors That Make a Difference

High-performing coaches:
- ☐ Listen carefully and attentively
- ☐ Are approachable and friendly
- ☐ Maintain an appropriate balance between talking and listening to others
- ☐ Express contrasting opinions in ways that encourage continuing dialogue
- ☐ Help others find their own answers rather than telling them what to do
- ☐ Are interested in what others have to say
- ☐ Work from the assumption that there is more than one way to solve a problem
- ☐ Employ questions to help others refine and improve upon their ideas and plans
- ☐ Encourage others to present their point of view—even when it differs from theirs
- ☐ Help others seek and identify learning opportunities in new assignments and projects

☐ Seek developmental feedback from others for their own professional development

☐ Encourage others to seek developmental feedback

DELEGATE AND INSTRUCT

Delegation is the fastest, most effective way of providing on-the-job developmental opportunities to others. Unfortunately, delegation is one of the most misunderstood and ineffectively used techniques in a manager's tool kit. In addition, the coach's responsibility of "instructing" or teaching a learner through a developmental assignment is also forgotten or ignored, further aggravating the situation. High-performing business coaches delegate often and effectively teach when the situation requires it.

Delegation is not "dumping" assignments or tasks on others, particularly when there is no developmental component associated with them. There must be something "new" (a challenge or stretch) in the delegated assignment; otherwise, it is merely passing on an assignment or task. High-performing coaches recognize that not all assignments are or will be developmental. But they do work hard to help others see the developmental opportunities that are present in many delegated tasks or projects.

High-performing coaches are also "situational delegators." They are acutely aware of the developmental needs of others and adjust their coaching strategy (personal time invested in planning, communicating and following up the delegated task) based on the complexity of the assignment and the degree of developmental stretch required by the individual responsible for delivering results.

The Behaviors That Make a Difference

High-performing coaches:
☐ Let go of the details when delegating
☐ Make expectations clear when asking others to do something
☐ Trust people to do a good job without constantly checking up on them
☐ Communicate information and instructions clearly
☐ Explain the purpose and importance when delegating assignments
☐ Review the success and failures of an assignment at completion
☐ Delegate challenging assignments that have developmental value

COMMUNICATE EXPECTATIONS AND MIRROR REALITY

Coaches must set clear goals and communicate (*verbally and nonverbally*) positive expectations/purpose when providing developmental opportunities to others. Providing frequent feedback to employees allows them to calibrate their progress and understand how others (*mirroring reality*) perceive them. Mirroring reality also requires the coach to continually provide the broader organizational context that helps others stay "connected" to the organization's strategy and values.

There are two ways in which successful coaches communicate their expectations. The "hard" side includes being clear when communicating the specific performance dimensions (what, where, when, quantity, quality, cost, etc.) of the developmental opportunity or assignment.

The "soft" side of communicating expectations is made up of all the unspoken behaviors (body language, eye contact, your accessibility, your degree of enthusiasm, etc.) as well as the types of assignments that you ask (or don't ask) others to do. These softer behaviors ultimately communicate the degree of confidence you have in others and your commitment to seeing that they have a successful experience (delivering on those "hard" results).

The Behaviors That Make a Difference

High-performing coaches:
- ☐ Provide others with timely feedback
- ☐ Avoid behaviors that may be threatening when giving negative feedback
- ☐ Focus others on the mission and goals of the team (department/organization)
- ☐ Provide a broad perspective, but leave others to determine the details
- ☐ Are generous in recognizing the contributions of others
- ☐ Take time to celebrate accomplishments when project milestones are reached

EXTEND SELF-RELIANCE AND BUILD NETWORKS

Helping employees become self-reliant (independent rather than dependent) is an important but intermediate step in the process of personal/professional development. This importance was demonstrated by the fact that *Promote Discovery and Self-Reliance* was its own competency cluster.

Successful coaches help others, once they are self-reliant, to see that sustaining the value of their perceived contribution over time is tied closely with their ability to collaborate and work with and through others (interdependent rather than independent).

The skill of building and extending networks is a critical element in the continuous process of learning and developing. Coaches that have strong and far-reaching networks can leverage their time through their personal and professional relationships with their colleagues so that they do not always have to be present to observe or coach others. High-performing coaches speed up the process of getting others visibility and connected to other parts of the organization.

The Behaviors That Make a Difference

High-performing coaches:

- ☐ Make sure important people in the organization recognize others' contributions
- ☐ Avoid taking personal credit for the accomplishments of others
- ☐ Encourage others to look for ways of helping (coaching) other individuals/groups
- ☐ Involve others from the work group with clients and customers
- ☐ Involve others in meetings or projects in other parts of the company

CONCLUSION

As we move into the next millenium, an organization's competitive advantage will still rest squarely on its ability to develop leadership capability quickly throughout the entire workforce. Instilling the attitude that continuous learning is everyone's responsibility and providing the behavioral descriptions demonstrated by high performing coaches to support the learning process can make a difference—for you, your peers/colleagues, your manager, and your organization!

ABOUT THE AUTHOR

Ron Cutadean is a principal with Novations Group, Inc. He has more than 20 years of experience as a management trainer, internal organizational consultant and human resource director in the utility, energy, aerospace and high technology industries. Most recently, he was the managing principal for Right Associates, Rocky Mountain Region. He has authored *Interviewing and the Selection Process* and is co-author of *Developing Leadership Capability: Coaching for Contribution*. Mr. Cutadean has a master's degree in economics from Duquesne University (1972).

Mixed-Sex Team Communication: Avoiding Negative Conflict

A. ECKSTAT

ABSTRACT

Sharing the same vision, understanding the same goals, and possessing the appropriate technical skills are all necessary for team success. However, these traits do not automatically result in a cohesive or effective mixed-sex work group or team. The ability to form an effective team is also dependent upon shared values and understandings and on the ability to communicate effectively within the group.

Substantive conflict within groups has been shown to have a positive effect; generally, it can enhance a group's performance. Affective conflict has been shown to have a negative effect; generally, it can retard a group's performance. Differences in the way men and women communicate can result in unintentional affective (negative) conflict, which will hinder the cohesiveness and performance of a mixed-sex team. Educating men and women about these sex differences, and establishing appropriate guidelines or codes of conduct, can facilitate more effective mixed-sex communication, avoid unintentional negative conflict, and reduce the gender barriers to a team's high performance. Examples of mixed-sex communication barriers and recommended guidelines to overcome these barriers are presented.

INTRODUCTION

The workplace continues to change. More organizations are forming teams to increase productivity. More and more, these teams are being made up of both men and women. Unfortunately, many of these same organizations are discovering that their mixed-sex teams are less cohesive, less productive and less effective than expected. The expected synergistic relationships often fail to form, and the results are often disappointing. One reason for the failure to meet these expectations is that communication problems exist between male and female team members.

A review of literature on teams reveals that effective communication between team

Arthur Eckstat, Doctoral Student, Nova Southeastern University, and Vice President, PBT Personal Bridges to Teamwork, Inc.

members is one of the requirements for a high-performance team (Amason 1996, Katzenbach & Smith 1993). A review of interpersonal communication literature reveals that men and women often attempt to communicate in different ways and have different meanings for and interpretations of the same words, phrases, intonations, and actions, in identical situations (Kramer 1977, Maltz & Borker 1982, Mendell 1996, Tannen 1990, 1994a, 1994b).

Substantive conflict within a group can be positive; it can enhance the group's performance (Amason 1996, Katzenbach & Smith 1993, Pelled 1996). Substantive conflict can encourage the positive exchange of ideas from the various perspectives of the team members. Conversely, affective conflict has been shown to be negative; it can reduce the group's performance (Amason 1996, Pelled 1996). Verbal affective conflict can be as blunt as a vulgar personal attack or as subtle as misunderstanding the intended meaning of a comment or utterance.

The ability to form an effective, high-performance work group or team is dependent upon shared values and understandings and the ability to communicate effectively within the group. Differences in the way men and women communicate can hinder communication effectiveness. Amason (1996) and Watson, Kumar, & Michaelson (1993) reported that process control improved the outcomes in diverse groups. Therefore, it is proposed that establishing appropriate guidelines or codes of conduct for mixed-sex work groups or teams can facilitate more effective mixed-sex communication, avoid unintentional affective conflict, and reduce the gender barriers to mixed-sex work group or team cohesiveness and effectiveness.

Effective communication within work groups and teams has become more difficult as women and men become larger fractions of the workforce in previously male-only or female-only and male-dominated or female-dominated occupations, respectively. In traditionally male-dominated industries, where the inclusion of women into traditionally male-only working groups is increasing, the resulting interpersonal communication difficulties can interfere with mixed-sex team effectiveness.

Following is a look at some of the differences between the sexes. The examples of male-female differences are stereotypical generalities. Individuals may be exceptions to the various differences cited. Taken as a whole, the author has found the examples to generally hold in more than 100 informal surveys taken between 1992 and 1997. Disputes among researchers with respect to why these differences between men or women exist (*nature or nurture*) or the degree to which they exist have not been considered. The examples of gender differences come from existing literature and the author's personal observations. The conclusions of how these examples affect mixed-sex work groups and teams and how the differences may be overcome are an extension of either the literature or the author's experiences. Structured tests of these conclusions are recommended.

Accompanying some of these differences are explanations of why some generally accepted team guidelines are especially effective for mixed-sex teams. Accompanying other differences are suggestions of ways for mixed-sex teams to avoid negative conflict through the incorporation of additional guidelines or rules of conduct. Educating men and women about these differences in the way men and women generally communicate may improve the performance of mixed-sex work groups or teams.

MALE-FEMALE COMMUNICATION DIFFERENCES

1. *Men tend to focus on solving a particular problem at hand, while women tend to see a larger picture.* One of the reasons brainstorming is an effective step in a structured Total Quality Management (TQM) process is that it provides men with the opportunity to do what women tend to do naturally: consider a larger picture. Instituting an *"Everyone Participates"* rule and including everyone in the group during brainstorming allows the more reticent team members of either sex to share their ideas. One of the keys to effective brainstorming is avoiding comments or discussion during the process so that no one needs to fear a negative reaction to their offering. Another key is to allow a word such as, "Pass" to be an acceptable response to anyone's turn to contribute. This provides for smooth flow where everyone can build on the contributions of others in a synergistic fashion. The use of "Pass" by everyone signals the end to the brainstorming exercise.

2. *Men tend to disseminate detailed information when presenting an idea. Women resent being lectured to.* Giving background details may be appropriate—and even appreciated—when men are communicating with other men. Too often, however, what a man sees as simply giving all of the necessary supporting information is interpreted by a woman as being a lecture. She may find it bothersome, if not offensive. Worse yet, a woman may interpret the telling of these details as a means of trying to persuade her to believe a weak argument or talking down to her. A team guideline to initially "Present Just the Idea," reserving details for later discussion or answering of questions, eases this mixed-sex communication problem. In addition, a *"No Speeches"* rule can act to trim the length of any presentation of viewpoints.

3. *Another facet of men's tendency to focus primarily on solving the immediate problem is the exclusion of alternative solutions once a satisfactory solution has been achieved. Women can often see how, and may just as often point out that, the team could have done better.* Men want appreciation for their accomplishments. They interpret a woman's remarks as to how the job could have been done better as lack of appreciation for their efforts, or worse yet, an attack on their abilities, often resulting in resentment and negative conflict. On the other hand, men like to move on once a problem is *solved*. They may not be readily open to the fact that there often is more than one right answer. Accompanying an observation that a better outcome may be advisable with a suggestion as to how, or asking if the team should consider some particular additional factor can open up a discussion without criticism.

   Time and budget restrictions may limit just how good of an outcome can be achieved. Early on, the team should *"Agree Upon the Team's Goals and Perform-*

ance Expectations" to help avoid gender-based conflicts and limit discussion once an acceptable outcome has been achieved.

4. *Men tend to interrupt others to make their point, while women tend to wait for their turn to speak.* The way a woman may judge such male interruptions ranges from supportive, at best, to impoliteness, or at worst, as a total disregard for her opinion or herself. Meanwhile, the offending men may only know that they are participating in a typical discussion in which it is important to present their views at the appropriate time or as the opportunity arises. Men or women can also use interruptions to dominate. A *"No Interruptions"* guideline for teams would level the playing field and avoid the interpersonal conflict arising from differences in what is considered normal or polite behavior. Formalizing the team process to the point of having to *"Raise Your Hand to Speak,"* and not speaking until recognized, will eliminate interruptions. Any participants can make short reminder notes for themselves if they fear forgetting their point before speaking. The degree of formality required may vary with the team or organization.

5. *Men tend to be direct, dictating to others and often seeking confrontation. Women tend to be indirect, negotiating differences and avoiding confrontation.* When working together, both men and women can benefit by using the phrase, *"Would You. . .?"* For men, using, "Would you. . .," in front of a request can remove the specter of appearing to dictate orders. For women, using, "Would you. . .," instead of, "Could you. . .," transforms the request, from a man's point of view, from questioning his *ability* to do something to determining his *willingness* to do something. Furthermore, a *"No Personal Attacks"* guideline can prevent unnecessary or unproductive confrontations. Even when a team member has failed to perform, the criticism should be focused on the activity instead of the individual. This will allow a continuation of the process without the attendant rancor of attacking a person's character or being.

6. *Men and women tend to differ in their views about giving or receiving unsolicited help.* Women will often help each other without being asked because it is a sign of caring between women. Men, however, may view unsolicited help as a sign that the person being helped is not trusted to perform on their own. This difference can result in unintended conflict and resentment in a mixed-sex situation. A woman may want unsolicited help when she *could obviously use it* or as recognition that she is being appreciated. She may resent being allowed to struggle alone by what she sees as uncaring or insensitive men. Meanwhile, the men, even if they want to help, may be avoiding insulting her by not trusting her to do it herself. The male experience leads them to expect her to ask for help if she wants it. A man may even think that a woman is just trying to prove she can do the job without asking a man for help. Conversely, a man receiving unsolicited help from a woman may either appreciate the help or resent her lack of trust in his abilities. Establishing guidelines to *"Ask For Help"* when you want it can avoid such conflicts. Making *"Would you like help*

with...?" a standard question for both sexes allows both the offer and the acceptance or refusal of the offer to be made without conflict.

7. *Both men and women may make inarticulate sounds during a discussion. However, the same sounds, such as "mm-hmm" or "uh-huh" generally do not mean the same things to men and women.* Men generally make these sounds to show agreement. Therefore, when a man hears a woman making *agreeable* sounds while he is talking, he may assume that she is agreeing with him. However, women generally use these same sounds to show that they are listening, not necessarily agreeing. A woman may make such sounds just to encourage further discussion. A man may think he has received agreement with his views from a woman only to hear her *change her mind* when talking to someone else or casting a vote. Conversely, a woman may be surprised to find that she has *taken a stand* on an issue when all she was doing was listening to an idea of one of her male coworkers. Formalizing consensus on ideas can avoid the effect of this difference in meaning for the same sounds or actions. *"Ask For Agreement, Do Not Assume It"* is a guideline that can avoid these mixed-sex misunderstandings over listening techniques or styles.

8. *Avoid the "What do you think?" pitfall.* When men ask, "What do you think?" they are generally asking for agreement or approval. When women ask, "What do you think?" they are generally asking for a discussion. A man may be surprised by an ensuing discussion with a woman that ends in agreement with what he originally said. A woman may be surprised by the shortness of a man's nod or comment and his quick departure to the next subject when she is looking for a discussion. Both men and women can benefit by being more specific in their requests for either agreement or further discussion.

9. *Women generally want to be liked and respected. Men generally want to be respected whether or not they are liked.* This is another facet of the fact that men tend to be more direct and confrontational while women tend to be more indirect and cooperative. This can also be a barrier to effective communication since it may lead to a subservient relationship between male and female team members rather than an equal teammate relationship. The guidelines presented here will help achieve that understanding of equality, thus enabling increased team cohesiveness.

CONCLUSION

The examples of male-female differences are stereotypical generalities. Taken as a whole, the author has informally found the examples to generally hold. The conclusions of how these male-female differences affect mixed-sex work groups and teams and how the differences may be overcome are an extension of either the literature or the author's experiences. Structured tests of these conclusions are recommended.

Negative conflict due to mixed-sex communication problems can be reduced in the workplace by incorporating the following guidelines or rules of conduct for mixed-sex work groups or teams:

- Everyone participates
- No interruptions
- Raise your hand to speak
- Agree upon the team's goals and performance expectations
- Present just the idea
- No speeches
- No personal attacks
- Ask for help
- "Would you like help with. . . ?"
- "Would you. . . ?"
- Ask for agreement, don't assume it

SUMMARY

Difference: Men tend to focus on solving a particular problem at hand, while women tend to see the larger picture.
  Guideline: *Everyone Participates*
Difference: Men tend to disseminate detailed information when presenting an idea. Women resent being "lectured to."
  Guidelines: *Present Just the Idea* and *No Speeches*
Difference: Men tend to exclude alternative solutions once a satisfactory solution has been achieved. Women can often see how the job could have been done better.
  Guideline: *Agree on the Team's Goals and Performance Expectations*
Difference: Men tend to interrupt others to make their point, while women tend to wait for their turn to speak.
  Guidelines: *No Interruptions* and *Raise Your Hand to Speak*
Difference: Men tend to be direct, dictating to others and often seeking confrontation. Women tend to be indirect, negotiating differences and evading confrontation.
  Guidelines: *No Personal Attacks* and use of *"Would you. . .?"*
Difference: Men and women tend to differ in their views about giving or receiving unsolicited help.
  Guidelines: *Ask for Help* and use of *"Would you like help with. . .?"*
Difference: In a discussion, the same inarticulate sounds, such as "mm-hmm" or "uh-huh," do not mean the same things to men and women.
  Guidelines: *"Ask for Agreement, Don't Assume It"*

REFERENCES

Amason, A. C., 1996. Distinguishing the effects of functional and dysfunctional conflict on strategic decision making: Resolving a paradox for top management teams. Academy of Management Journal, 39: 123–148.

Katzenbach, J. R., & Smith, D. K., 1993. The wisdom of teams: Creating the high-performance organization. Boston, MA: Harvard Business School Press.

Kramer, C., 1977. Perceptions of female and male speech. Language and Speech, 20: 151–161.

Maltz, D. N., and Borker, R. A., 1982. A cultural approach to male-female miscommunication. Language and social identity, ed. By J. J. Gumperz, 196–216. Cambridge: Cambridge University Press.

Mendell, A., 1996. How men think: The seven essential rules for making it in a man's world. New York: Random House. Paperback: Ballantine

Pelled, L. H., 1996. Demographic diversity, conflict, and work group outcomes: An intervening process theory. Organization Science, 7: 615–631.

Tannen, D., 1990. You just don't understand: Women and men in conversation. New York: William Morrow. Paperback: Ballantine.

Tannen, D., 1994a. Gender and discourse. New York: Oxford University Press.

Tannen, D., 1994b. Talking from 9 to 5: How women's and men's conversational styles affect who gets heard, who gets credit, and what gets done at work. New York: William Morrow.

Watson, W. E., Kumar, K., and Michaelsen, L. K., 1993. Cultural diversity's impact on interaction process and performance: Comparing homogeneous and diverse task groups. Academy of Management Journal, 36: 590–602.

ABOUT THE AUTHOR

Arthur Eckstat is vice president of training and organization development at PBT Personal Bridges to Teamwork, Inc., in Phoenix, Arizona. He has more than 25 years of successful experience in personnel development and project management, including Total Quality Management (TQM). He has more than 10 years of experience facilitating human potential seminars and workshops.

Mr. Eckstat received a bachelor's degree in mechanical engineering from Detroit Institute of Technology (1968) and a master's degree in business administration degree from the University of Phoenix (1985). He is currently working on his doctoral dissertation on organizational mixed-sex conflict, in pursuit of a doctoral degree in business administration at Nova Southeastern University.

Mr. Eckstat is a member of the Academy of Management and the National Association of Gender Diversity Training (NAGDT). In August 1997, he was honored as a contributor to aerospace history during a ceremony held at the Smithsonian National Air and Space Museum in Washington, D.C.

Mr. Eckstat can be reached at PBT Personal Bridges to Teamwork, Inc., 2632 East Mountain View Road, Phoenix, Arizona 85028; by e-mail at PBTeckstat@ uswest.net.

Dealing with Office Politics

W. C. EGGERS

ABSTRACT

"Office politics" sounds like a very negative subject. To an engineer who teams with others it can bring up images of unfair and annoying—if not devastating—work environments. In this paper we will examine what office politics is and what it is perceived to be. We will look at individuals' success and what is sometimes looked at as their using "backdoor techniques" to achieve success, when in reality they may just be good communicators and smart in the ways they present their creative work. To the average engineer, some of these tips may appear to be bold and less than desirable because they are not grounded in technical facts or because they lack sound principles of physical reality. However, techniques that improve our ability to relate to other humans cannot be discounted. In the final analysis, the success of our professional product will be based on the judgment of other human beings.

POLITICS—WHAT IS IT?

Politics in the broad sense, is a positive term. Influencing others to do something we see as good, impacting others in order to make progress, representing the collective will of the people in a democratic process; these can all represent positive actions. Because of their creative nature, engineers are particularly sensitive to creative work that is "their own." Justifiable reward should follow good creative work and more creativity should follow without impedance.

Our country was founded by people such as Benjamin Franklin and Thomas Jefferson, who recognized the value of creative work. We need only to compare the patent laws of the United States, where inventors have a legal right to reap the rewards from technological breakthroughs, to other countries' laws to find that technology and its progress are highly valued by our democratic system. Transfer these rights to an employer as we do when we accept employment and we see how we have the traditional right to receive

William C. Eggers, P.E., Management Consultant

81

credit for our work in the office or workplace. Let someone subvert that right by taking credit for our work or by drawing away rightful recognition or by diverting us from the resources to do creative work and we have an emotional situation and a justifiable disdain for actions that can be called politics. Someone has been influenced in a negative sense. We can see that there are good and bad politics—not in the sense of placing judgment on how slickly (good, but bad) the political action was carried out, but on what the intent was.

POLITICS—SO WHAT?

The image of an entrenched elected political figure bowing and serving special interest groups, taking money illegally and not serving the will of the people is an onerous one. Is this what happens in the work place? Maybe the answer is "no," but the label—and especially the negative label—of "politician" is appropriate for all the reasons that ethical people can define. The office politician can bring teamwork and productivity to a halt. The fear that a product that you have produced might be misapplied or the concern that someone else might gain credit for your intellectual work can lead you to a frustrating existence.

In most organizations much time is wasted by negative office politics. Recently a large temporary staffing service conducted a nationwide survey. Assuming it was an objective study, it concluded that executives spend 20 percent of their time or one day a week dealing with office politics. That loss in productivity can be measured in dollars. As an example, suppose a vice president making $250,000 a year wastes 20 percent of his salary; this sounds like enough money to be concerned.

POLITICS—CAN WE DO ANYTHING?

If we narrow our focus to "bad" office politics, the kind that subverts an organization or undermines the work of people, the self-promoting plotting that others do, then yes, there are things that can be done. Suggested remedies might include direct confrontation with the person who is undermining you. This may be awkward but it can prevent the growth of a negative situation. Document and date your work so you can show ownership. Never engage in self-serving or idle criticism of coworkers or management, which can lead to undermining your integrity. Know your employer. If you find the self-promoting politician is getting rewarded, consider counseling with your boss. If that is not successful, consider another job. As previously stated, you have the right to be rewarded for your own work; this idea is basic to our democratic traditions.

EFFECTIVE COMMUNICATIONS—GOOD POLITICS?

For purposes of what follows, let us define good politics as the ability to influence others to think or act in a way that we find mutually beneficial. Good politics starts with the ability to communicate. It is important, then, to examine communications in terms of influencing and transferring information. It is important here to know where we are going with this. We want to establish rapport (as a positive political condition) with others. Rapport exists when a common trust and interest exists between two people. Positive communications begins with our own self-awareness. Such self-awareness can be broken down into four stages: 1) Unconscious Incompetence; 2) Conscious Incompetence; 3) Conscious Competence; and 4) Unconscious Competence. Take the example of riding a bicycle. In the *Unconscious Incompetent stage*, if we have never ridden a bike we can admit that we have no clue as to what to do or how to do it. In the *Conscious Incompetent stage*, we know there must be something we can learn in order to handle a bike. In the *Conscious Competent stage*, we can learn a series of steps such as sitting, pedaling, balancing with the handlebars and maintaining the necessary speed. Finally, we come to the *Unconscious Competent stage* and we don't have to "think" about how to ride; we carry that knowledge with us forever. This is how good communication techniques become part of us and how we are able to influence others "as second nature," as the term goes.

Having an awareness of how we experience events in our life is important to reaching an improved communication level. The components of experience are: 1) Internal Experience; 2) External Behavior; and 3) Internal Computation. *Internal Experience* is the part of us that reacts to things in an emotional manner—how we feel about something. The *External Behavior* is the "doing" or the actions we take with our body—what the world sees. The final experience is the *Internal Computation*—the thinking process. Be aware that we all receive stimuli within these three frames of reference.

What are the types of communications we need to be aware of in order to improve our communication ability? *Intrapersonal* communication is the communication that takes place in our own mind. This is the starting point for good communications. Positive self-talk, self-esteem, confidence, and winning attitude all relate to a sound basis for good communications with others. Knowing yourself, liking yourself and doing positive things for yourself are all part of intra-personal communications. All should be accomplished with an ethical respect for others. Here is where we can see a negative example of the bad office politician who perhaps likes himself or herself too much. *Interpersonal* communication is the one-on-one form of communication. Interpersonal communication cannot be effective unless it is done in a state of rapport—the positive political condition we strive for. *Group* communication is the process of relaying information to several people at the same time. This is probably easier than one-on-one communication because it involves less interpersonal dynamics. We only have to think about the great lecturers or entertainers who lack personal conversational skills to illustrate this difference.

The following communications principles could be argued about, but if they are accepted for now, they can provide a basis for improvement.

1. *People respond to their own reality.* We all receive information within the framework of what we understand as reality. Getting to understand others' reality provides us with a reference point to begin communication.

2. *Anything can be accomplished if it is broken down into small enough chunks.* No job is too complex. (Notice that it was not said too time-consuming, but rather that it cannot be broken down into small, doable tasks.) For example, the Apollo moon landings were accomplished with a series of small steps.

3. *The value of your communications is the response you get.* Communication elicits a response from the individual with whom you are communicating. If you don't get the reaction you expect, you need to be flexible and shift your communication form.

4. *The person who can be most flexible will control a situation.* This may take the form of taking a modified position on a subject and thereby gaining insight into the other person's understanding of reality. Arguing a technological application on the basis of technology alone to a person who is a financial expert will reap little understanding until you can shift and discuss the economic advantage of applying the technology.

5. *There is no such thing as a failure; there are only opportunities for feedback.* This is illustrated again by the example of trying to convince a financial expert that an application of a certain technology is a good idea by relying on a technical discussion only. If we understand that our temporary setback was not based on the merits of the technology, which by the way, might be our own reality as an engineer, but by our failure to communicate the financial reality of the situation, then we have used feedback in a positive way. At this point we should be starting to understand good communications.

Perceptual filters that our target individual has are important to how we communicate. If we gain awareness of the other person's unconscious behavior, we could increase our ability to be flexible. Think of the boss who always finishes our presentation in his or her terms. So unconscious is his or her behavior that he/she is unaware that the track you were on may have been leading to a different place. Understand the built-in conclusion the boss has and you can present the information in a manner that will successfully bring you to the place you want.

We need to understand how well we are communicating—while we are communicating. We should be aware of the recipient's behavior as a way of establishing how effectively we are communicating. Sensory acuity is an awareness of the subtle unconscious behaviors that are clues to how we are being received. For example:

1. *Extremities*—notice crossing and uncrossing of legs, fidgeting with fingers, tapping of feet; these are signs that all is not well; rapport is not ours yet.

2. *Body Movements*—shifting of the body, movement or tilt of the head; these are signs of our communication's effectiveness.

3. *Facial Expressions*—flaring of nostrils, lower lip movement, eye patterns; the face is the first place where anger or non-approval is expressed.
4. *Breathing*—a sure sign of positive or negative excitement, and something to watch.
5. *Voice Patterns*—change in pitch or volume.

Some people have been known to display complete opposites of the usual behavior. For example, think of the boss who speaks in a loud volume manner but is perfectly calm inside. When that boss' voice drops to a level that most polite people use, you know trouble could be brewing.

Effective communicators are also aware of the representation that words have and how each of us usually prefers one pattern of word communication to another. Consider the person who is a *visual communicator*. Words such as aim, clear, perceive, envision, and viewpoint will dominate their conversation. *Auditory representations* are made with words such as amplify, call, verbalize, orchestration, and harmony. *Kinesthetic representations*, or feeling words, are feel, skew, fumble, probe, and grasp. We need to listen to the language of the person with whom we are communicating and frame our own communication in words that fit their preferred representation system: visual, auditory, or kinesthetic.

The goal of all of this awareness is to help us improve the effectiveness of our own communications by establishing rapport. As stated before, rapport is essential to good communications. If we want to drive meaningful results in our work teams and compete with bad office politicians, we now have some tools. Rapport is ours to establish with others, if we become aware of ourselves in a positive manner; understand and value the unconscious behavior of others; watch for feedback signs that are the products of unconscious behavior; be aware that word patterns used by others are the most comfortable communication forms they could receive; and understand a person's perception biases, and know their own reality, which is the only reality we must deal with if we are trying to communicate with them. If we frame our communications with some of these simple principles, then we have taken a major step in establishing rapport and communicating with and influencing others.

PLANNING—GOOD POLITICS?

Anytime we work with others and we attempt to influence the outcome of events, we must start with a plan. Good politics or the exertion of positive influence starts with knowing what we want to do, when we want to do it, and what resources we have. Good planning starts with letting ourselves know how to do it before we try to convince others.

In the discussion about communications we emphasized the importance of words and the frame of reference they symbolize. The use of words is also very important in planning. The first step in making a plan our own and one that represents what we want should include *stating goals in positive terms*. For example, say "I want to integrate these systems," not "I don't want this equipment incompatibility anymore." *Goals should be initiated and controlled by you*: "I have the budget available," not "I should get more money." *First steps should be sensory- based*: "I will spend four hours a week reading in the cool and quiet comfort of my home," not "I need to read more." *First steps include appropriate chunk size*: "I will design this bolt pattern on the gusset," not "I am going to design a bridge." A mental frame of reference to establish effective communications with others begins with communicating with ourselves intrapersonally.

Basic to good planning is a sense of realistic thinking, not pie-in-the sky thinking; it is a reasonable and doable series of interrelated steps to meet a goal. There are a number of considerations involved in good planning:

- *What do you want?* This definition to yourself should clear any vagueness in thinking.
- *How will you know when you have it?* Define it in terms of quantifiable parameters.
- *Where, when, and with whom, do you want it?* Know the others involved. This may be an entire customer base or one person, such as your boss.
- *How will the desired outcome affect other aspects of my life?* If your goal is more responsibility and more money, will you consider a move to another city?
- *What stops you from having your outcome already?* Are the barriers real or perceived?
- *What resources do you have?* Resources include people and things.
- *What additional resources do you need* ? This may be a critical consideration, since you will now involve other people and will identify the possible expenditure of somebody's money.

The real worth of a plan is the commitment to actions. Look for a leader who will champion your outcomes. It is necessary to involve the centers of power in a work environment. We are reminded here that good rapport with a leader is essential; this is positive politics. Do you have buy-in from others? Identify the stakeholders and establish that rapport. Cultivate a group of people who know what you want and are supportive.

Do you have an understanding of the effect of this change? Maybe a new process you are proposing will eliminate part of someone else's work while at the same time opening new opportunities for them. It may be time to identify the positive advantages to the others and gain their support. Again, this is good politics rather than the bad politics of the "cut-throat operator."

Will your sponsor mobilize others? If you have established rapport with the right sponsor and have laid out your plan, you should count on that sponsor to communicate to others. This is the nature of good politics.

Do your sponsors understand the link to other systems? You chose the sponsor because he or she has a more global view of your business. This should give you the opportunity to find out more about your business and the linkages and ripple effect of your plan. Have you clarified first steps to others involved? Just as the end of any business meeting should include a definition of next actions, clarification of what you are going to do and what you expect of others is important to good planning.

Supporting and keeping your plan alive and well requires that you identify who is opposed to what you want to do and who is supportive. This could take the form of a written list, identifying who supports your plan and opposes it as well as why they support or oppose it. A good idea might be to identify each person's level of support or opposition in terms of high, medium or low. You can then work on an action plan to maintain support or gain support from those who were initially opposed. Mobilize commitment in terms of who will be responsible for what and when. This is an opportunity to define expectations and bring an action plan down to simple tasks that must be accomplished to bring about your desired outcome. The final action to be done is to develop a checklist of what has to happen when, and who is responsible. The checklist, if intelligently done, will give you the tool to keep the plan going.

SUMMARY

Good politics practiced to influence others for a positive good is something that can be learned. Working in an environment where the self-serving politician disturbs others with unethical behavior should be avoided. Good communications applied to gain rapport with others, along with sound planning, will go a long way toward combating the negative effects of people who want to compete for the limelight or personal financial gain at the expense of others.

We have looked at a learning process of good communications by examining how humans process information and how their own unconscious filters are to be understood so that we can achieve rapport with the people who affect our lives. We have looked at some basic planning principles that lead to getting good things accomplished by identifying what, when, who, and with what tasks need to be linked. Along with planning, the important element of gaining others' buy-in and sponsorship is most important. No matter how elegant the technical solution, it will go nowhere without other people agreeing and supporting it. Agreement and support can be gained by using good communication techniques and good planning; both are elements of good politics.

BIBLIOGRAPHY

Becker Associates, Boulder, Colorado, *Communicating For Maximum Performance,* Workshop at MDC Learning Center, July 1997.

D. Ulrich, *Executive Development Series,* MDC Learning Center, May 1997.

ABOUT THE AUTHOR

William Eggers is a management consultant with a strong background as an engineering manager. He is a leader in career development, skills assessment and training of engineers and their managers. His most recent, noteworthy corporate assignment was with McDonnell Douglas, now Boeing, managing the concept definition, development and construction of a Corporate Learning Center in St. Louis, Missouri. Working with the CEO and other corporate executives he converted their vision of a Learning Center into reality.

Mr. Eggers began his career as an electronics engineer and progressed to chief engineer, working in the areas of airborne radar development, automatic test equipment development, simulator and training equipment development, and spacecraft system integration. These assignments have included line management and program management responsibilities.

Mr. Eggers is a registered professional engineer in the state of Missouri. He has a bachelor's degree in electrical engineering from Washington University in St. Louis, Missouri, and a master's degree in management from Maryville University. He is a member of the IEEE-USA Career Maintenance and Development Committee and an Associate Fellow of American Institute of Aeronautics and Astronautics.

Mr. Eggers can be contacted at 328 Northmoor Drive; Ballwin, Missouri 63011; telephone (314) 391-8013; e-mail wmeggers@swbell.net.

IEEE Engineering Management Society: A Resource for Your IEEE Responsibilities and Your Career

G. GAYNOR and C. VOEGTLI

ABSTRACT

Ask any engineer why he or she belongs to IEEE and the answer will usually be, "I need to stay technically current in my area of expertise" or "I need to stay abreast of the latest developments in other technical areas." Ask anyone why he participates in local IEEE activities, and you get the above answers, in addition to "It's a great opportunity to interact with colleagues who have similar interests."

These responses also hold true for members of the IEEE Engineering Management Society (EMS). A common theme among active EMS members is the belief that, for companies to be competitive, executives and managers must be just as concerned as individual contributors are about staying current in their roles. Great technology and products are not enough; companies must manage their endeavors creatively and effectively in order to ensure success.

More insight into the EMS comes from members' past experiences, such as one member's account of an EMS chapter meeting in Silicon Valley: "The speaker, a vice president of engineering at a local company, delivered an interesting presentation on how his company runs and rewards project teams. As stimulating as the presentation was, I was actually even more energized from the question-and-answer session that followed the presentation. The questions asked were questions I received day-to-day as an engineering manager and project manager. The audience's dialogue with the speaker gave me some great ideas for solving specific problems at my company. I also realized that I had found valuable resources for the future: the expertise and insights of the speakers and audience members at EMS events.

Today EMS chapters worldwide are dedicated to bringing these resources to their members and to the management community at large. IEEE is mainly known for its technically focused societies, but each technical area also is influenced by engineering management issues. And if you are an IEEE volunteer, your activities most likely involve some management as well: managing events or meetings; planning programs; and motivating others to help. The management knowledge and

Gerard Gaynor, Past President, IEEE Engineering Management Society
Cinda Voegtli, President, IEEE Engineering Management Society

skills members learn through EMS are just as applicable to volunteer work as they are to our paid careers.

This article describes the benefits of membership and participation in the EMS and the details of specific activities offered by its chapters. These resources can provide you with assistance as you execute any IEEE volunteer responsibilities, as well as support for your overall career.

EMS MEMBERSHIP BENEFITS

The EMS currently provides benefits in four primary areas:

- Publications
- Education
- Awareness
- Networking

PUBLICATIONS

Publications have always been a keystone benefit of EMS membership. Members receive the official EMS publications: *Engineering Management Review* and *Transactions on Engineering Management.*

Engineering Management Review is a compendium of key articles reprinted from leading management publications around the world, including *Harvard Business Review, California Management Review, Research and Technology Management, the Journal of Product Innovation Management*, among others. The articles are selected for their immediate value to practitioners in industry. Issues focus on themes, such as project management, the virtual enterprise, quality, change and competitiveness, human resources, innovation, and global partnerships. The *EMR* also includes a review of great management books each quarter.

Transactions on Engineering Management is a more research-oriented publication, with regular issues containing articles in five categories:

- *Strategic and Policy Issues* include authoritative discussions on strategic directions of technology management.
- *Research Articles* present the results of ongoing or completed research.
- *Technical Management Notes* comment on papers previously published and discuss experiences using published results.
- *Focus on Practice* papers describe implementation problems and present solutions with significant implications.
- *Book Reviews* are critical evaluations of recent books on engineering and technology management.

Special issues marked the 40th anniversary of the journal and covered concurrent engineering; information technologies and the transfer and commercialization of technology; international R&D projects; and Total Quality Management in engineering management.

EDUCATION

Education should be a life-long process for engineering management professionals, just as it is for any technical individual contributor today. It is also one of the keys to maintaining an effective organization in today's fast-paced environment. One form of education occurs when we learn about new areas, such as when we gain cross-functional knowledge and increase our professional breadth by learning adjacent or parallel processes. We see this when manufacturing processes are adopted by engineering or when hardware development procedures are adopted for software development. However, we also learn when we view a known methodology with a different focus, such as using a theoretical or ideal viewpoint instead of a practical or "real" one. This technique can provide a new, different, and potentially valuable perspective for our daily work processes.

The primary mission of EMS chapters is to provide focused, practical, accessible educational opportunities for the engineering management community in their areas as well as colleagues in other functional areas. They seek to highlight both proven techniques used at successful companies and new, leading-edge management theories and techniques. Chapters generally provide this education by arranging two types of programs: monthly evening presentations on topical issues, and periodic seminars on subjects of value to the management community. The subjects often covered by EMS meetings fall into four categories:

- Engineering and Project Management
- Business and Program Management
- Management Skills
- Personal Professional and Career Development

Our primary audience consists of line managers, technical leads, project and program managers, executives, and engineers interested in becoming managers. Engineering and business executives benefit both by gaining personal knowledge in their areas of interest and by having their management and technical teams well educated in these areas. EMS monthly programs and seminars attract a wide range of engineers and managers from high-tech businesses, including engineering managers, directors, and vice presidents; project and program managers; and sales and marketing managers and executives.

These programs, offered by the Silicon Valley guest speakers, are typical of the variety of topics available through local EMS chapters:

- Software Measurements for Product and Process Improvement
- Commitment-Based Planning
- Persuading Your Audience
- How to Hire and Retain Technical Experts
- Consistently Effective New Product Development
- Practical Project Management
- Software Project Management
- How to Run a Small Company
- Venture Capital in Santa Clara Valley

Many EMS chapters hold joint meetings with other IEEE societies. For instance, one EMS chapter held a meeting with the local IEEE Computer Society chapter on intellectual property issues in software development and management.

The EMS also sponsors seminars to provide more in-depth education, with an emphasis on practical management techniques and tools. Case studies and group exercises are used to provide immediate hands-on application of the theories and techniques presented. Past seminar programs by EMS chapters have included:

- Managing Your Way to More Competitive Product Development
- Quality Rapid Product Development
- Breakthrough Project Management
- Keys to a Winning Product Introduction
- Behavioral Interviewing
- Managing Effective Meetings
- Negotiating Skills for Project Managers and Engineers
- Time Management for Engineers
- How to Start Your Own High-Tech Business

EMS aims to be a management resource throughout IEEE. You can gain access to seminars like these through your local EMS chapter, or your section or other IEEE society can contact EMS to help you sponsor such a seminar.

The EMS as a society is also involved in conferences that focus on management issues. For instance, the EMS sponsors an annual International Engineering Management Conference, which includes tutorials and presentations made by management professionals from around the world.

The EMS is also a cooperating society for the ProjectWorld Conference, which is held each year in the United States and is now offered in Europe. This conference brings together technical and project/program management practitioners from around the country (and now the world) to discuss management issues. The session tracks cover such subjects as: Implementing Project Management, Leadership and Team Building, Project Start-up Preparation, Strategies for Coping with Super-Urgent Demands, Reducing Time-to-Market, Project Management in Software Design, Business Process Re-engineering, and Selecting Project Management Software.

EMS also offers management tracks, workshops, or papers at other IEEE conferences. If you are involved in another society or in conferences put on by your section, EMS would be happy to provide speakers to address management and professional development topics.

AWARENESS

Awareness of what's new in "management technology" is related to education but is slightly different. It is really a precursor to education. Our industry is changing rapidly. There are new management tools and new management processes to match the changing technology. Every day we hear more about virtual corporations, partnering, alliances, self-managed teams, telecommuting, work groups, etc. Managers must have an awareness of these changes and what they mean to them before they can mount an effort to become current by educating themselves.

One way to become aware of new trends is to listen to the thoughts, opinions, and predictions of respected industry and academic leaders. These are the people chapters invite to speak at their monthly meetings. Topics of interest to engineering managers are selected by the chapter's officers and volunteers, based on feedback received from members and attendees of previous EMS events. The speakers sought for these subjects are executives, business people, entrepreneurs, financial experts and/or authors. For instance, past speakers in Silicon Valley over the last few years have included Gordon Bell, Regis McKenna, Bob Pease (National Semiconductor), Marv Patterson (Director of Corporate Engineering at Hewlett-Packard), Marjorie Balazs (President, Balazs Analytical Laboratory), Josef Berger (Vice President of Advanced Development, Tencor Instruments), John Glavin (Vice President of Engineering, Conner Peripherals), Michael Murphy (General Partner, California Technology Stock Letter), Curt Wozniak (Vice President of Engineering, Sun Microsystems) and Jim Forquer (Vice President of Worldwide Operations, Apple Computer). Speakers typically make their presentation and answer audience questions for an hour, then make themselves available following the meeting for one-on-one conversations. The meetings provide the opportunity to crystallize and enhance any new ideas in discussions with the speaker and with your peers and associates.

NETWORKING

Networking is recognized by outplacement organizations as the single most effective mechanism for finding a new job. However, this is only one aspect of its value. Networking is a way to share information within the management community about techniques, processes, policies, and methodologies that can be used to address problems we all experience. Networking can provide a cost-effective way to exchange information in the area of management technology with friends, peers, industry leaders, and new contacts. Simply stated, it's a valuable way to avoid "reinventing the wheel." The time before and after EMS meetings and seminars, and during team exercises and breaks at the seminars, is prime time for making new contacts and sharing management issues and expertise.

CHAPTER OPERATIONS

The Engineering Management Society was formed within IEEE in 1954. Since then, more than 40 EMS chapters have been created around the world, with more forming each year. Local EMS members form a chapter by having 12 EMS members sign a petition and get approval from their IEEE Section.

Chapter organizations are run by volunteers from the local management community. These volunteers plan and implement the chapter's monthly programs and seminars. Chapters are structured and operated according to IEEE chapter guidelines. Officers (Chairman, Vice Chairman, Secretary, Treasurer, and representative to the Society Board of Governors) are elected every spring. Several other positions are often appointed each year as well: program chair, seminars chair, membership chair, and publicity chair. Other volunteers serve as "project managers" of individual EMS program events.

The chapter executive committee, composed of the elected officers and other volunteers, meets periodically to set the chapter direction and plan upcoming events. Some chapters receive important input from informal industry advisory boards made up of executives from local technology companies.

Chapter program meetings are generally held from two to 10 times per year, either during lunchtime or after work. Some chapters sponsor seminars during the year. Attendance at workshop-style seminars is typically limited to 15 to 25 people, to encourage effective interaction; more lecture-oriented seminars may be attended by 50 to 70 people. All EMS chapter events are announced in the local IEEE section publication.

THE FUTURE OF EMS

We in electrical engineering and related fields are at the "bleeding edge." Hardware is cheap and it's the age of software, virtual reality, multimedia, the "information highway," CASE tools, software measurement, ISO-9000, and life-cycle planning. Management must grow and adjust to meet the challenges of telecommuting, virtual corporations, new aggressive business structures, and global competition. EMS is working to help managers meet these challenges. We welcome your input and participation:

- What new creative management ideas do you have? What contributions do you have to make to the quality of high-technology management?

- What speakers do you want to hear?
- What do managers today need in their toolbox?
- What other programs and services would be of value to you and other managers?

THE RELEVANCE OF EMS

Information about management is important to functional or line managers, project managers, program managers, executives, engineers in technical leadership roles, engineers interested in becoming managers, consultants, and anyone else who wants to understand the process of getting things done. Participating in the EMS is a valuable means for you to keep in touch with the mainstream of information about the management of technology and the development of high-tech products, as well as leadership and management skills that will help you in any personal or professional endeavor. EMS participants read about and discuss recent events and anticipate changes in the industry, are exposed to expert viewpoints, and entertain and challenge new ideas in a peer environment, away from the driven rush of daily business. We seek answers to the challenging management questions that we all face. Through participation in the EMS and its programs, we invite you to have an impact on your project, your company, your IEEE work, and your own career.

ABOUT THE AUTHORS

Gerard (Gus) Gaynor is immediate past president of the IEEE Engineering Management Society. He is also Editor-in-Chief of *Today's Engineer*, a new magazine that focuses on the non-technical aspects of an engineer's career. He is a retired 3M executive and is currently President of G.H. Gaynor and Associates, a management consulting firm in Minneapolis.

Cinda Voegtli is current president of the IEEE Engineering Management Society. She is president of Emprend LLC, a project management consulting and publishing firm in Silicon Valley, and is a consulting partner with Global Brain Inc., a consulting firm that focuses on techniques for rapid product development. Previously she held director-level engineering management and senior project management positions at several high-technology product development companies in California and Texas. Ms. Voegtli holds a bachelor's degree in electrical engineering.

IEEE ENGINEERING MANAGEMENT SOCIETY

More than *two-thirds* of engineers spend more than *two-thirds* of their careers in managing technology, product development, and engineering-related projects and activities.

The Engineering Management Society addresses their needs, providing access to the latest advances and "best practices" in management.

Who joins the Engineering Management Society?

- Industry and government engineers and technical, program and project managers and executives
- Engineering and management researchers, faculty, and students
- Members of affiliated technical and management organizations

Members are involved in engineering and management of:

hardware, software, systems, and processes
design, testing, manufacturing, and maintenance
projects and programs
technology planning and transfer
basic and applied research.....

...For the resources you need to manage your engineering and technology-based endeavors

...For the most up-to-date and broad-based coverage of management research and practice available

...For a powerful network of peers worldwide

MANAGERS AND ENGINEERS:
Join the Leaders
of Your Profession!

The *Transactions on Engineering Management:* A research publication rated as best technology and innovation management journal.

The *Engineering Management Review:* A reprint journal of the best recent articles on engineering, technology, and project management from such respected publications as the *Harvard Business Review.*

The *Engineering Management Newsletter:* EMS news, events, and articles of interest to members.

Educational resources: A wide range of management and technical conferences, seminars, products, and books and publications at special member rates.

Local chapters worldwide: Local events with presentations by respected experts and practicing managers and executives, for continuing professional and career development.

MEMBERSHIP OPTIONS:

IEEE/EMS Membership:	*EMS Affiliate Membership:*	*Student Membership:*
Members of IEEE and EMS receive: • All EMS benefits outlined above ***Plus:*** • Low member rates for all IEEE conferences • *IEEE Spectrum* and *The INSTITUTE* • IEEE Financial Advantage Program (insurance, mutual fund, credit card, annuity and banking service programs, professional liability coverage, and various business discounts.)	**EMS Affiliate Members receive:** • All EMS benefits outlined above Admission as an EMS Affiliate requires membership in another scientific, technical, or management society OR the endorsement of three IEEE members.	**Student Members of IEEE/EMS receive:** • All EMS benefits noted above ***Plus:*** • Low student rates for all IEEE technical and management conferences, symposia, and workshops • *IEEE Spectrum* and *The INSTITUTE* • IEEE Financial Advantage Program

Basics of Intellectual Property: Patents, Copyrights and Trademarks

A. H. GESS

ABSTRACT

Knowledge is power. It can also be intellectual property when it belongs exclusively to one person. Intellectual property has taken on a much more significant role in global economics since World War II. This is due in a large part to the shift away from the traditional industries—cars, steel, mining, and textiles—to high-technology information-based industries. Industries based primarily on knowledge and fast-rising technologies, such as semiconductors, advanced materials, biotechnology and information technology, are the ones experiencing economic growth today.

The law of intellectual property provides a competitive edge to these growing industries. Patent, copyright and trademark laws are the tools of intellectual property. What these tools are and how they can be used to the inventors' advantage is as important to today's fast rising technology-based industries as the technology that fuels their growth.

INTRODUCTION

The value of intellectual property lies in the owner's ability to prevent others from exploiting it. In civilized countries this is done in the courts. The U.S. legislative scheme for protecting intellectual property reflects a policy decision to leave the economic valuation of intellectual property to the marketplace.

The unparalleled material and cultural benefits America enjoys today is due largely to the United States' patent and copyright laws. The incentive furnished by these laws to inventors, authors, and composers has fostered a flood of creativity that has enriched society. Patent and copyright laws encourage inventors and authors to continue their efforts by virtue of the rewards garnered to them as the result of the rights granted by these laws. Trademark laws protect the public goodwill created by the products of this creativity.

Albin H. Gess, Founding Partner, Price, Gess & Ubell

The framers of the U.S. Constitution were students of history. They knew about earlier industrialized periods of the world's history and prior governments' use of a patent system to encourage technology and new industry. The Venetians had a patent system that protected the rights of an originator in 1297. In 1331 England, the King issued letters of protection to industrialists who would settle in England for the purpose of starting a new textile industry. In 1629, the English Statute of Monopolies was enacted.

The U.S. Constitution (Article 1, Section 8) is the basis of United States intellectual property law. Intellectual property is a monopoly mandated by statute, or common law, enabling the owner to prevent others from using, making or selling the protected invention, creative work or trademark. Monopolies allow owners to exclude competing products or processes from the market, which translates into economic power. The basic tools of intellectual property protection are patents, copyrights, and trademarks.

PATENTS

The underlying concept in granting patents is that greater innovation and technological growth is encouraged by rewarding inventors for the uncertain and expensive process of research and development. Because a patent requires public disclosure of the invention, other companies may use the disclosed information freely when the patent expires, or with permission during the patent term. It is also thought that patents can help avoid costly research duplication because the information they contain can be used and distributed. Moreover, patent disclosure is thought to spur new technologies and to improve technical efficiencies in related industries.

U.S. patent law has its source in the U.S. Constitution, which empowers Congress to "promote the progress of science and useful arts, by securing for limited times to . . . inventors the exclusive right to their respective . . . discoveries." In response to this directive, Congress passed the first patent statute in 1790. The original patent statute has remained relatively stable over the years; the last major revision was made in 1953. Most of the revisions made to this statute have responded to directions (or misdirections) taken by the courts. Recent international trade agreements such as GATT are driving major changes to the U.S. patent law.

Patents are the best known form of intellectual property. This may be due to the press coverage given to large patent damage awards in such cases as *Honeywell v. Minolta*, ($140 million); *Polaroid v. Eastman-Kodak*, ($873 million); and to individuals like William Kearns, who sued the automotive industry on his intermittent windshield wiper patent and has collected $20 million to date.

THE PATENT APPLICATION

Patents are granted by the U.S. federal government after an examination process.

The exclusive rights granted in a patent are for 20 years from the application filing date. Patents involving compositions or processes subject to approval under the Food, Drug and Cosmetic Act may be extended for the period during which they were subject to review by the FDA before marketing, for up to five years.

The patent statute provides that "Whoever invents or discovers any new and useful process, machine, manufacture, or composition of matter, or any new and useful improvement thereof, may obtain a patent therefor, subject to the conditions and requirements of this title." The simplicity of these words has led to many court cases. Their meaning seems to change as society's attitudes change and science comes up with new and never-before-imagined inventions. The courts have concluded that naturally occurring articles or processes—things that nature has invented—cannot be patented. Fundamental ideas or laws of nature cannot be patented.

The general categories of patentable subject matter are: (1) process, (2) machine, (3) article of manufacture, (4) composition of matter, (5) any improvement in (1)–(4), (6) new and asexually reproduced plants, and (7) any new and ornamental design for an article. What do we mean by "process?" A process is a method, and it includes a new use of a known process, a new use of a machine, a new use of an article of manufacture, a new use of a composition of matter, or a new use of a material.

An application for patent must include: (1) a specification; (2) a drawing; (3) an oath by the applicant in the form prescribed by law; and most important, (4) a filing fee, which changes consistently upward.

The *specification* is a written description of the invention and of the manner and process of making and using it, in such full, clear, concise, and exact terms as to enable any person skilled in the art to which the invention pertains or to which it is most nearly connected to make and use it. The specification must describe the best mode contemplated by the inventor to carry out his invention. This means you must describe your invention in its best embodiment at the time you file for patent application. You cannot hold back a more efficient structure, disclosing only a less-efficient, more expensive or cumbersome structure. The specification must also include one or more claims that particularly point out and distinctly claim the subject matter that the applicant regards as his invention. Claims are strange to most laypersons and require considerable skill to draft. A claim, like a poem, places emphasis on every word and the interplay of words. Grammatically, a claim is the object of a simple sentence. A claim may be several lines long or several pages long.

A drawing or drawings must be provided when a drawing will help illustrate the invention and facilitate the understanding of the subject matter sought to be patented. The only time a drawing is not required is when the invention is something you cannot illustrate by a drawing, such as a chemical compound that is best described by a chemical formula.

The Commissioner of Patents may sometimes require a working model of the invention or require the inventor to furnish a specimen or the ingredients for the composition of matter, for experimentation purposes. According to current practice, models are required only when the Examiner does not believe that what the inventor described will work. Something that does not work for its intended purpose is not useful, and only new and useful inventions are patentable. An example of a device that is held in such dispute is a perpetual motion machine, a machine that puts out more energy than it takes in.

In the United States, with minor exception, an application for patent must be made by the inventor. The applicant must swear that he believes himself to be the original and first inventor of the process, machine, article of manufacture, composition of matter, or improvements thereof for which he asks for patent. Whenever an inventor refuses to sign an application for patent or he cannot be found, the person to whom the inventor has assigned the application, or has agreed in writing to assign the invention, or who has a proprietary interest in the invention, may make application for patent as an agent for the inventor. This means your employer can apply for patent in your name without your signature, if he convinces the Patent Office that what you invented belongs to him.

EXAMINATION PROCESS

So, all you have to do is write up your invention, describing its most efficient operating embodiment, draft some claims, provide some drawings, and swear that you are the original and first inventor, and you get a patent? Not so fast! The statute says "...subject to the conditions and requirements of this title." Applicants must jump over some hurdles before they are granted a patent. These conditions and requirements are listed in two important sections of the patent law: 35 U.S.C. § 102 and 35 U.S.C. § 103. Section 102 describes at least seven hurdles the applicant must clear. Section 103 says even if you clear all the hurdles of § 102, you still cannot have a patent, if the difference between what you are trying to patent and what somebody else has already described or patented would have been obvious to a person of ordinary skill in the art at the time you made your invention.

The 35 U.S.C. § 103 standard is subjective and, perhaps, the most difficult barrier. Typically, a Patent Examiner assumes the position of a person of ordinary skill in the art. He or she will combine two, three, or more prior art references to reject the application. Prior art can be U.S. or foreign patents, published articles, or textbooks that predate the application. The Examiner will conclude that, in light of these combined prior art references, the invention is obvious and therefore not patentable. As one might surmise, this creates a lot of argument.

For the most part, with the guidance of a patent attorney, patents are granted eventually. On average, for electrical/electronics patents, the process currently takes about three years from filing.

CASHING IN

Once you have your patent, what do you do with it? You can sell it, lease it, or wave it in the air like a club to ward off your competition. If you sell it, you take your money and do something else. If you lease it (patent license) or wave your patent around, you must be prepared to deal with infringers. Infringers are those who make, use, or sell your patented invention in the United States without authority during the term of your patent; those who actively induce others to infringe your patent; or those who sell an essential part of your patented invention. How do you deal with them?

I always recommend talking first. Ask the infringer to stop; ask them to pay money for the damage they have caused you; or ask them to agree to pay you money under mutually acceptable license terms. If they do not want to deal, your remedy is a civil action in federal court for infringement of your patent. You come into court with your patent presumed valid. A variety of defenses may be asserted against it. The most common defense is "I don't infringe." Another defense is that the patent is unenforceable because of some bad things the inventor did. Or, the defendant may attack the patent by claiming it is invalid because the Examiner made a mistake in granting the patent in the first place.

But you prevail. What can you get? You can ask for and get an injunction, an order from the court commanding the infringer to stop what he is doing. You can ask for and get money damages. The amount may be computed on the basis of a reasonable royalty, along with interest and costs, or upon another basis or theory, such as lost profits. You can ask the court to increase the damage award. If the court feels that the infringer was willful and deliberately violated your patent, the court can triple the damage award and can award you all your attorney's fees on top of it. Now we're talking millions.

COPYRIGHT

Copyright is a unique form of protection issued by the Library of Congress (Copyright Office). Most people are familiar with the copyright symbol ©, which often appears on the title page of a book. This symbol provides an indication to the reader that the author, whether the actual writer or the publisher, is claiming a proprietary right in the expression of the ideas contained within the book.

Copyrights, like patents, spring from our Constitution, Article 1, Section 8. A desire to harmonize intellectual property law throughout the world led the United States to become a signatory to the Berne Convention. As a practical matter, this altered formalistic requirements under U.S. Copyright Law, the most visible requirement being that it is no longer necessary that a copyright notice be specifically carried on the work. The presumption that a work is available and free to the public, if it does not have a copyright notice, no longer exists. A copyright notice still carries value, since statutory damages are not awarded under U.S. law unless there is a copyright notice.

REQUIREMENTS OF COPYRIGHT

A prerequisite for protection of a work under copyright is that the work must be fixed in a tangible medium of expression. Digitized information that is machine-readable is considered a tangible medium of expression. For example, an extemporaneous speech would not be copyrightable unless it was recorded. A baseball game would not be subject to a copyright unless it was recorded. An impromptu or improvised musical composition would not be subject to a copyright unless it was recorded.

Another requirement is that the work must have a minimal amount of originality. The originality required is directed more to independent creation and does not require, for example, the novelty of a patent. Originality basically means that the work owes its origin to the author and is not copied from other works. Thus, two people can take a photograph of the same identical object and both have created an original work.

PROTECTION PROVIDED

The scope of copyright owners' protection is limited to preventing others from *copying their work*. If a person independently produces a similar or even identical work, the copyright owner has no claim against the independently created work. To prove a case of copyright infringement, a copyright owner must establish both a substantial similarity between the two works and actual copying, or a set of circumstances that would enable a reasonable interpretation that only copying could have occurred.

It is important to place the copyright notice, preferably a "©", the author's name, and the date of first publication, on the work. Pursuant to the Berne Convention, such a notice removes any defense of innocent infringement and enables the securing of a federal copyright registration. Prior to 1978, failure to provide the copyright notice in the United States could have resulted in the work being dedicated to the public.

Registration of a copyright is necessary to qualify for statutory damages and attorneys' fees. If an infringement began immediately after a work was published and the registration occurs within three months after the first publication of a work, or if there is a registration prior to the commencement of the infringement, statutory damages and attorneys' fees are available. (One court has basically stated the significance of registration is "merely the plaintiff's ticket to Court," since the actual protection of the copyright arises at the time of the creation of the work.) Another value of a copyright registration is that the U.S. Customs Service will bar the importation of copies of a registered work.

REGISTRATION

The Copyright Office, which is a section of the Library of Congress, divides copyrights into different groups and provides appropriate forms for the different categories:

- Class TX (non-dramatic literary works);
- Class PA (works of the performing arts, such a musical works, dramatic works, pantomimes and choreographic works, and motion picture and other audiovisual works);

- Class VA (works of visual arts, including all pictorial, graphic, and sculptural works); and
- Class SR (sound recordings and any recorded literary, dramatic, and musical work embodied on the same record that contains the sound recording for which the registration is sought).

The primary function of the Copyright Office is to process copyright registration forms. When copyright registrations are issued—usually four to eight weeks after you file your request—you receive certain substantive rights, such as the ability to sue for actual and statutory damages in a federal court and the ability to request attorney's fees for intentional infringement. The cost of securing a copyright is $20.00. The Copyright Examiner will look at the forms and may raise formal and even substantive objections by letter. You may communicate by telephone and letter with the Copyright Examiner to resolve the objections.

The Copyright Office takes the position that if the sole intrinsic function of an article is its utility, the fact that it may be uniquely and attractively shaped will not qualify it as a work of art. If, however, the utilitarian article incorporates features—such as artistic sculpture, carvings, or a pictorial representation—that can be identified separately and are capable of existing independently as works of art, the object can be subject to registration as to those features. Applying this standard is complicated. For example, an ornamental wire wheel for a car that is purely decorative and does not serve any support function was still said to be a utilitarian article that would prevent the lug nuts, brakes, wheels, and axles from corroding. In another example, an ornamental drawer pull was held unprotectable as a utilitarian object, even though there were sculptural aspects to its design.

AUTHORSHIP AND OWNERSHIP

The issue of who can file for a copyright is sometimes important, especially in an employer/employee relationship. The Copyright Act refers to a "work made for hire," which consists of a work prepared by an employee within the scope of his or her employment, or of certain works specially ordered or commissioned. With regard to the latter, the U.S. Supreme Court has said that a hired party is to be considered an employee under the general common law of agency where the hiring party controls the manner and the means by which the product is produced, as opposed to simply the right to the final form of the product. Thus, a sculptor who is hired to create a sculpture of a football player and is paid for the sculpture would not be divested of his right to the copyright of that sculpture. This may be a surprise to the owner of the sculpture. The owner commissioning such a work should carry the relationship one step further and require the assignment of any copyright as a condition of employment. For our purposes, anyone who is commissioning an outside person to do work (e.g., a computer programmer to produce a program) should have a written agreement that says the employer is the owner of the copyright in all works produced. Likewise, it would be prudent in any employment agreement form to also have a general clause that indicates the ownership of proprietary rights, including any copyrights. Under the law, literary efforts of an employee remain the property of the employee, if they are not rendered within the scope of his or her employment.

A written agreement relating to the transfer of a copyright in a work made for hire requires the signature of both parties, not simply the person who performs the work. The agreement should be signed prior to completion of the work. An agreement made after completion of the work cannot create a work made for hire retroactively. This artificial distinction is important because the life or term of a copyright will vary depending upon authorship—namely, whether the work is pursuant to an anonymous work or work made for hire, or whether it is the original creation of the author. At present, a copyright term begins at the creation of the work and endures for the life of the author plus 50 years after his death. If there are multiple contributors to the work, the time period starts from the death of the last surviving author. With regard to a copyright based on a work made for hire, the term of the copyright extends for a period of either 75 years from the date of first publication of the work or 100 years from its actual creation, whichever expires first. The term "publication" has a specific legal meaning, but for our purposes simply refers to the distribution or availability to the public.

EVERYDAY ACTS OF INFRINGEMENT

We have all used a copying machine in a library to make copies of certain copyrighted publications or portions of text. A specific section in the Copyright Law exempts this act. In libraries that have a public collection, members of the public can access the books and make a single copy of a publication. If you ever made multiple copies of a publication at a library, you fell outside the exemption and violated the copyright of the publication.

Most of us have VCRs and use them to tape copyrighted transmissions or broadcasts from cable or television. The ability to make such copies has been much debated and litigated. The issue became academic upon passage of the Audio Home Recording Act of 1992. This Act puts the burden on the manufacturers of digital audio recorders to establish a fund to compensate recording artists and copyright owners for any purported lost revenue that would occur from home taping. At the same time, a grant of immunity was given to the individual who makes copies without any direct or indirect commercial motivation; that is, when he or she makes copies for his or her own use. We are, in essence, paying for the right to make copies as part of the purchase price of the VCR.

The basic concepts of copyright protection are not altered by using the Internet. But the ability to hold people legally responsible, when publications can be sent almost instantaneously with a large distribution around the world, is presenting a considerable problem to copyright owners of large commercial interests. For example, Walt Disney has pressured America Online, Inc. and other commercial on-line services to either remove images of its characters or to display only authorized reproductions.

It has been estimated that there are more than 60,000 computer bulletin board systems. There is an on-going debate over who is responsible for the matter posted on such bulletin boards. At least two federal courts have held bulletin board operators liable for on-line copyright infringement.

TRADEMARKS

Most people recognize that the name of a product or service has value to consumers as an indicator of a product's worth and value to the product's owner in terms of reputation. Thus, it follows that unknown "Brand X" is often inexpensive and that well-known name brands frequently bear "department store prices." Furthermore, most people recognize that a manufacturer does not want someone to "rip off" his well-known brand name and use it to sell cheap copies of products.

But why does a well-known brand name cost more? Why do owners of brand names get upset when their brand names are used without authorization? Exactly why is a manufacturer so sensitive about incorrect use of his product's brand name? For example, why is a person always supposed to make photocopies instead of "xeroxes," and why is it important to apply a BAND-AID-brand adhesive bandage rather than a band aid? With a little background concerning trademarks and service marks, you can understand some of the ins and outs of this somewhat complex legal area and you can help any business with which you become associated to enjoy adequate protection for the names of its products.

TRADEMARK ORIGINS

You probably already know that a trademark is the legally protected name of a product. Similarly, a service mark is essentially the same thing as a trademark, but it is used for a service. Trademarks are a bit more complicated than just being simple product names. The rules that dictate whether or not a given name can act as a valid trademark are somewhat mysterious to the uninitiated.

Although trademarks are now firmly a part of "intellectual property," their origins are somewhat different than those of copyrights and patents. In their oldest and simplest form, trademarks appeared as a mark or symbol that a craftsman or manufacturer placed on his goods. Such marks appear on goods produced as long ago as several millennia before the birth of Christ. It is not uncommon to find marks on jewelry and stone work from Mesopotamia and Ancient Egypt. Both the Greeks and Romans used distinctive "brand marks" on the seals of amphorae of wine.

PROTECTION FOR THE PUBLIC

As modern capitalism developed during the Renaissance and following periods, many businesses began to place some sort of trademark on their goods. The motivation for doing so was probably pride in workmanship. Moreover, trademarks represented a particular business' products to its customers.

Today, it is generally recognized that trademarks have at least two important functions. First, they protect the public from poor-quality goods by ensuring that a given product can be recognized readily. Second, the manufacturer who owns a particular trademark is given a means to protect the public goodwill he builds for products through efforts to maintain quality and through advertising.

PROTECTION FOR THE BUSINESSES

Manufacturers of high-quality goods develop goodwill or an anticipation of quality from their loyal consumers, who associate a given trademark with a certain type and quality of product. Almost everyone expects that the brand-name products will be more expensive. This is partially because the manufacturer has spent money improving the quality of his product and seeks to recoup that expenditure. It is also partially because brand-name products are usually more heavily advertised, and that expense must also be recouped. Lastly, the manufacturer charges more because he knows that the market will bear a higher price. That is, many consumers are willing to pay a premium to eliminate the uncertainty that surrounds purchase of an unknown product.

A trademark (service mark) is anything that can be used to distinguish the goods or services (service mark) of one provider from those of another reliably. This will often be a product name but, harkening back to trademarks' historical ancestors, often includes a logo or some other such identifying mark. This means that a good trademark will be distinctive of the goods. Conversely, a poor trademark will be incapable of distinguishing the goods. This incapability may be either inherent in the mark or may be due to the fact that the mark in question can be confused with the mark of some other manufacturer, and thus fails to be distinctive. Do not confuse trademarks with trade names, which are simply the names that businesses use. Some trade names become trademarks, but only if they are used to distinguish the goods of the business.

PICKING A TRADEMARK

In essence, our legal system is set up to protect "good" trademarks. This protection serves the trademark owner because it protects the customer goodwill that the trademark owner has established by maintaining good quality and by advertising. This protection serves the consumer by making it possible to purchase products of known quality rapidly and reliably. Generally, legal protection in the United States comes from two sources: state law and federal law. Most of the basic details of trademark law have developed under the common law (court-made law), which has been reinforced by state statutes. More recently, there has been an overlay of federal trademark law, which serves to reinforce and consolidate (and in some cases extend) the common law.

Trademarks vary considerably in their inherent ability to distinguish goods or services. The most distinctive class of marks is called "arbitrary" (or "fanciful") marks. These marks are often made up words or logos. Because the words are new and without any connotation, they have the ability to instantly distinguish goods. An example of an arbitrary mark would be KODAK. Today, this mark has almost universal identification with photographic products. But when George Eastman coined the word, it had no meaning whatsoever.

A second group of marks that is capable of distinguishing goods is known as "suggestive." These marks suggest some aspect of the goods and often involve some sort of word play. Thus, once the consumers figure out the gimmick involved, they will always remember the name. A good example is the trademark WINK for a grapefruit-flavored soft drink. Merely hearing the term WINK does not describe the product. However, when you know that grapefruit is involved, it is pretty easy to remember getting grapefruit juice sprayed in your eye while cutting a ripe grapefruit. Thus, the suggestive implication is that the product tastes so much like fresh grapefruit that your eyes will wink shut expecting a squirt of grapefruit juice.

The next step down is a "descriptive" mark—one that tends to describe some aspect of the product. Most novices at creating trademarks tend to come up with descriptive marks, albeit with a funny spelling. Thus, someone will want to name their new charcoal lighter "CHARKOALLITER." A classic type of descriptive mark is the use of a surname in the mark: "SMITH'S PORK & BEANS." This type of mark is not capable of primarily distinguishing a product. However, if such a mark is used over a period of time, consumers may come to recognize it as distinguishing a particular product. At that time, the mark is said to have acquired "secondary meaning." The idea here is that although the mark cannot have a primary meaning capable of distinguishing a product because it describes the product, through repeated use and familiarity the mark can gain the ability to distinguish a product. The take-home lesson is to eschew cute misspellings that are primarily descriptive of a product and instead go for suggestive or arbitrary terms.

FEDERAL REGISTRATION

A federal trademark act was passed in 1946. Called the Lanham Act, it has been amended several times since it was passed and forms the basis of federal trademark law. The Act specifically preserves all state and common law rights. It brings uniformity to the law of trademarks by providing a nationwide registry for all marks. It also confers several special advantages to those who use it. First, a federal registration allows a trademark owner to sue an alleged infringer in federal court without any of the traditional jurisdiction tests. Second, federal registration provides *constructive notice* relating back to the date that that the application was filed. Once a trademark is registered, everyone is deemed to have notice of the mark. Constructive notice gives a trademark priority over everyone except those who have used the mark (or a similar mark) or have applied for registration prior to the date of filing. Therefore, no one else can legally adopt that mark as their own or use a similar mark that is likely "to cause confusion, or to cause mistake, or to deceive." Once a mark is registered federally, no one can establish rights to the mark by using it in a different geographical area than the registered owner. However, since the Lanham Act preserves common law rights, if the mark is already being used in two separate geographical areas at the time of registration, the owner of the registration has rights throughout the whole country *except* for those regions where the prior user had already established state or common law rights.

ABOUT THE AUTHOR

Albin H. Gess is a founding partner in the intellectual property law firm of Price, Gess & Ubell in Irvine, California. He has been a practicing patent attorney since 1972. He is a senior member of IEEE, and is a member of the California Bar (1972), U.S. Supreme Court (1977), and U.S. Court of Appeals for the Federal Circuit (1982), and is registered to practice before the U.S. Patent and Trademark Office (1972).

Mr. Gess received a bachelor's degree in electrical engineering from the University of Detroit (1966) and a juris doctorate from the Washington College of Law, American University (1971). He is a past section chair of the Orange County Section of IEEE (1995, 1996); and was member-at-large of the L.A. Council (1996). He is currently treasurer of the Orange County Computer Society chapter and the Orange County Communications Society chapter. He served as president of the Orange County Patent Law Association (1989) and as vice president of the Orange County Engineering Council (1996).

Mr. Gess can be reached at (949) 261-8433 or by e-mail at a.gess@ieee.org.

Communicating with Your Customers: Getting into Your Customers' Worlds to Serve Them Better

J. HARPER

ABSTRACT

This workshop presents communication skills critical to building positive relationships with customers. Businesses need to step into their customers' worlds and relate to them on a human as well as a business level. In today's fast-paced business environment of rapidly changing market needs, it is becoming increasingly difficult to cultivate personal relationships with customers. The attitude of "what have you done for me lately" requires that businesses not only meet their customers' needs but anticipate them as well. Customers are lost when they perceive that businesses are more concerned with their own agendas than with customer satisfaction. In most cases, the root cause of this negative perception is poor communication. Customers want to be shown they are cared about. Unfortunately, they often learn to believe otherwise. Businesses need to develop a true sense of purpose in order to step into their customers' worlds and relate to them on a human as well as a business level.

Interactions should be positive experiences for both the customer and the business; therefore, it is imperative that customers feel both valued and respected. Communicating this message should be relatively easy. However, in situations that involve dispute, taking a defensive position is often the first reaction on both sides. Damage to the customer-supplier relationship is a frequent result, but confrontations need not result in losing customers. With good communication skills, a positive relationship can be built and maintained.

In this paper, readers can learn how to:

- listen to understand
- view the world through their customers' eyes
- gather and give information at a customer's level of understanding without appearing arrogant or patronizing
- empathize with a customer without seeming phony

Using these skills will result in a more open and empathetic relationship, which will enable the supplier to anticipate the customer's requirements better. Suppliers build

John Harper, Vice President, Training, PBT Personal Bridges to Teamwork, Inc.

credibility by demonstrating insight into their customers' worlds. Customers come to realize that the business both listens to and cares about them. The net result is consistently superior service based on recognizing and understanding the customer's true needs.

INTRODUCTION

Customers want to continue doing business with suppliers they believe care about them. They want to feel that the supplier is working with them. In fact, customers will often "bend over backward" to maintain such business relationships. Caring about the customer is all about addressing the personal or human issues that result from business interactions. This is viewed as being so basic that it is often overlooked. Suppliers often assume that their customers will be satisfied as long as the goods or services are provided.

Most people think their customers are only those who purchase goods and services from their business. These are actually *external* customers. *Internal* customers, on the other hand, are those people within the organization who are rarely recognized as customers but are, instead, called "co-workers." Communication with these internal customers is generally considered a given. However, leaving internal customer relations to chance often leads to negative results that adversely affect productivity and morale. For the purpose of this paper, a customer is defined as anyone inside or outside the company who receives goods or services from someone in the company. Having this mindset is critical to improving communication with all customers.

The most successful businesses address both personal and business issues when dealing with their customers. Not only are the desired goods and services identified, but the human (emotional and personal) issues are addressed as well by using appropriate communication skills. Identifying a customer's business issue requires problem-solving skills and approaches that are quite different from those used to address emotional issues. Root causes of such business issues as late delivery of product can trigger emotional outbursts. The emotions involved must be recognized and addressed before problem solving can be initiated. Ignoring the emotions may create the perception that the supplier does not care about the customer but rather cares only about doing business and making money. The perception of not caring is a primary reason for customers leaving a supplier. The business that recognizes customers as fellow human beings with emotions that must be addressed as a part of doing business is in a much stronger position to keep its customers than one that focuses only on business issues.

Skills to be discussed in this paper are:

1. Active Listening—listening to understand
2. Empathy—recognizing emotions and situations
3. Obtaining Customer Input—encouraging open and honest communication

COMMUNICATING WITH YOUR CUSTOMER

"They don't listen and they don't care" are the words usually spoken by a customer about to leave you for your competition. There is no silver bullet for improving or salvaging relationships with customers. Rather, a combination of business skills (e.g., Total Quality, follow up, scheduling, etc.) and *soft* skills (e.g., communication, social graces, mixed-sex communication, interpersonal, etc.) must be applied appropriately. Some of the most important soft skills are active listening, empathy, and obtaining customer input.

ACTIVE LISTENING

The most important skill for dealing with customers is active listening, since it results in understanding. "You heard me, but you weren't listening." This strong statement is heard every day in one form or another: "You're not listening" or, "That's not what I said" or, most dramatically, by nothing more than silence. Of all the feedback, silence is the worst response you can receive, because it indicates clearly that the other person has given up trying to communicate; they've come to the conclusion that further efforts would be energy and emotion wasted on a lost cause.

Active listening is defined as the skill of listening to understand. This means that you, as the listener, must BE QUIET! You work hard to make sure the image you have created in your mind is identical to that which the speaker has in their mind. An active listener must keep the following points in mind:

- Do not make judgments of or pre-judge what the speaker is saying. Remember that you are listening to understand.
- Avoid the temptation to formulate rebuttal arguments or to state your position while the other person is talking. Keep your mind open to what is being said.
- When the person has finished making a point, feed back what you understood the person said by paraphrasing in your own words. For example, "So what you are telling me is . . . (paraphrase)," or "You're saying that . . . (paraphrase). Is that right?"
- It may be okay to interrupt for clarification or to validate your understanding. One form you can use to interrupt is, "Let me make sure I understand what you just said. You're saying . . . (paraphrase). Is that right?"
- Before you both proceed with the discussion, make sure the speaker agrees with your feedback. It may require several iterations to get it right. This is, in fact, asking for permission to continue the dialogue.

Taking the time to understand the other person shows that you respect them and value what they are saying. Otherwise, why take the time? It also encourages the other person to speak more freely, which is especially important in resolving business and emotional issues. In a debate you may be able to use the other person's statements to strengthen your own position. You may also learn something that will make you rethink your position.

Active listening is a powerful skill that requires discipline to master. Properly used, it not only increases understanding, it encourages the other person to practice it as well. This is a form of leading by example.

EMPATHY

For our purposes, empathy can be defined as recognizing the customer's situation and the emotions that are associated with it. You do not have to agree with the emotions or the situation. You just have to recognize and validate them. When the emotions have been addressed, you can begin to work the business issues. With emotions in the way, business issues become much more difficult, if not impossible, to be resolved.

Many times, people just want to talk about their feelings. This is often called *blowing off steam* or *venting*. When a person is venting, it's best to keep quiet and let them talk. This is a good time to use active listening skills to understand why the person is experiencing such strong emotions.

Feedback, or comments that validate—and often de-fuse—the emotions and situations include:

- "That sounds like a difficult situation."
- "If that happened to me, I know I would be upset."
- "I can see why you would be upset."

Even simple comments like, "Wow," accompanied by appropriate body language such as shaking your head, shows the other person that you recognize and appreciate their position and the resulting emotions. When you recognize that emotions have been calmed, you should direct the discussion toward correcting the specific problems that caused the situation. Avoid prolonged discussion of the emotional aspects of the problem (sympathy) and move on to problem solving. This moves the focus from the problem to the solution.

Statements that redirect the discussion from emotion to problem solving usually begin with one of the five "W" words or the one "H" word: who, what, when, where, why, or how. For example:

- "What do you think caused that?"
- "Who was involved?"
- "Where did it happen?"
- "Why do you say that?"
- "How can we correct this?"

Should the person want to continue to vent or stay in their emotions, go back to empathetic comments and gestures and repeat the process. Remember that you do not have to agree with the other person's emotions and situation. In fact, you may disagree. What is important is that you recognize their position and acknowledge their right to it. The important thing is to deal with the other person's emotional state before directing the discussion to resolving the business issues.

OBTAINING CUSTOMER INPUT

One of the biggest mistakes businesses make with customers is to assume they know more about what the customer needs than the customer does. The tendency is to offer solutions before the customer explains their problems or requirements.

Asking the right questions is two-thirds of solving the problem. The objective is to understand the customer's situation and their needs. This requires asking the right questions and *listening* to the response. It is imperative that you understand the situation from the customer's perspective. Understanding begins with getting the big picture by using open-ended questions and requests to encourage the flow of information. Some examples are:

- "What is the problem?"
- "Explain what happened."
- "Tell me what went wrong."
- "How did it affect your...?"

Once the big picture becomes clear, closed-ended, limited-scope questions can be used to verify understanding and obtain confirmation or clarification. For example:

- "As I understand you, what happened is . . . (paraphrase). Is that correct?"
- "How many times did it occur?"
- "What time should we deliver?"
- "Will you pick it up here?"
- "Is there anything else?"

Using a combination of open- and closed-ended questions will get you the information you need efficiently and accurately. Use open-ended requests as much as possible, because they encourage the customer to volunteer information; you are asking them for their input rather than *interrogating* them with a series of closed-ended questions. Few things are as frustrating as wanting to talk about an issue but being prevented from doing so by a (perceived) know-it-all who performs a quick interrogation and then proceeds to tell you what he or she sees as your problem and your solution. No matter how well intentioned your actions as the business representative may be, the risk is that the customer may perceive you as arrogant if you do not give the customer representative the opportunity to tell his or her story and be understood. You run the risk of the customer thinking "They can't listen and they don't care."

SUMMARY

Improving your relationships with your customers is a continuing process that requires the use of human relations as well as business-related skills. Knowledge of the business, services, and products that are provided is essential. However, the business with superior human relations skills will be more successful because the customer will feel listened to and cared about.

Active listening, empathizing with the customer, and obtaining customer input are three powerful human relations skills. Appropriate use of these skills results in insights into the customer's world as seen through their eyes. When a business makes the effort to look at a situation from the customer's perspective, the customer receives the message that the organization they are doing business with appreciates their situation, listens to them, and cares about their satisfaction.

Of course, the most important factor in applying these skills is demonstrating a sincere desire to understand your customer. To do less is to risk being perceived as phony and manipulative—a sure-fire way to inflame the customer beyond redemption. A genuine interest in the customer's position, on the other hand, will bring in and maintain loyal customers who will want to stay with you for the long run.

ABOUT THE AUTHOR

John Harper is vice president of training at PBT Personal Bridges to Teamwork, Inc., in Phoenix, Arizona. He has more than 30 years of experience in personnel development and program management. He previously served as chief engineer and as program manager in a Fortune 100 company. He has implemented and facilitated training programs for all organizational levels in the areas of communications, customer relations, supervision, teamwork, and Total Quality Management.

Mr. Harper received a bachelor's degree in mechanical engineering from City College of New York (1962), is a registered professional engineer, and has been granted several patents. Several nationally recognized training programs have certified him as an instructor and for train-the-trainer instruction.

Mr. Harper also mentored minority students for the Arizona State University Career Services Department S.O.L.I.D. (Student Opportunities for Leadership, Internship and Development) Program, where he facilitated student projects designed to teach Total Quality techniques and give students a competitive edge in the business world.

For more information, contact Mr. Harper at PBT Personal Bridges to Teamwork, Inc., 2632 E. Mountain View Road, Phoenix, Arizona 85028; telephone (602) 788-3083.

Life Work: Reducing Stress and Planning Your Career Toward Personal Values

J. A. HOSCHETTE

ABSTRACT

To be successful, engineers and scientists must have a plan for dealing with stress. The best way to reduce stress in one's career is to develop a career mission statement and strategic plan using personal values as a foundation. This paper will help engineers take a snapshot of where they are in their careers and will provide guidance and direction for career planning.

INTRODUCTION

Most professional engineers and scientists go through several career crises over the years, for different reasons. Some experience career stress late in their careers, while others experience it just a few years after graduating from college. The most common cause is the lack of progress professionally. They realize their careers have not progressed as fast as they had hoped. They left school with an idealistic vision of engineering and ambitious career plans. They set high goals for themselves, but years later they realize they are not "making it." Some call this being "stressed out" or "burned out."

Another reason for career stress in engineers is frustration with their jobs due to poor assignments or lack of good technical challenges. Jobs are often more mundane than expected, and engineers may find themselves solving difficult "people" problems along with the technical ones they are assigned.

Still another cause of career stress is lack of recognition. One of the most career-crushing realizations, after successfully completing a momentous project, is not to be recognized for the great sacrifice of personal time and effort that contributed to the effort's success.

These are only a few of the career issues that arise for engineers, causing them to ask, "Am I where I want to be?" Questioning your job is good and natural; it will happen often throughout your career. In order to answer correctly, you will need a career mission

John A. Hoschette, Member, IEEE-USA Career Maintenance and Development Committee, PACE Chair, Santa Clara Valley CA Section

statement, goals, and a strategic plan. Your answer to this question will be yes if you are fulfilling your career goals and are happy—only then will you have a low stress level.

DEVELOPING A CAREER MISSION STATEMENT, GOALS, AND A STRATEGIC PLAN

The process for developing a career mission statement, goals, and a strategic plan is shown in Figure 1. This process consists of four key steps: self-evaluation/identification of principles, values, and goals; writing a mission statement; generating a career plan; and, finally, taking action. Are career plans necessary? They are if you plan to fulfill your dreams of what an engineering position should be. Succinctly stated, failing to plan is simply planning to fail.

Self-evaluation is the first step in the process. Self-evaluation consists of taking an inventory of your personal beliefs and principles. Once you have determined what values and principles are most important to you, the next step is to formulate a mission statement based on them. The third step is to generate a career plan, and this is followed by action!

Because engineers want their career plans and mission statements to be perfect, they often put too much emphasis on the product and fail to realize that the process is equally important.

SELF-EVALUATION AND IDENTIFICATION OF VALUES

An excellent means of determining your values in life is given by Stephen Covey in his book *The 7 Habits of Highly Effective People.* Covey says that we all have a set of personal principles or values that govern our actions. These are shown in Figure 2, along with a short list of work principles. The latter are a subset of our personal principles and determine our actions at work.

The self-evaluation process starts with determining the priority of the personal principles in our lives and then the priority of our work principles. You should rank these principles from the most to the least important. To help understand these principles,

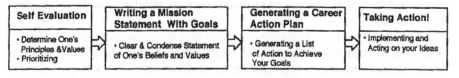

Going Through The Process Is As Important as The Product !

FIGURE 1. Process of developing career mission statement, goals and strategic action plan.

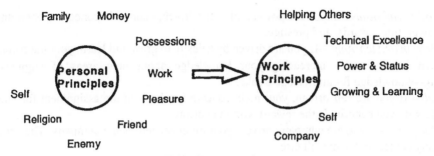

FIGURE 2. Principles.

let's briefly review the definition of each and how they drive our decisions. Starting with the personal principles, in no particular order of importance:

Possessions—If you are possession driven, your main concern in life is to obtain such possessions as cars, clothes, or houses. You measure your value by the number of your possessions.

Self—If you are self-driven, your main concern is what's in it for you. How much do you get out of it?

Money—If you are money driven, your main concern is how much money you have and how fast you can get more.

Spouse—If you are spouse driven, your main concern is what you can do for your spouse to make his or her life better.

Religion—If you are religion driven, your main concern is living your life according to the principles of your religion and observing religious rules.

Enemy—If you are enemy driven, your main concern is getting even or taking revenge on your enemies.

Family—If you are family driven, your main concern is family activities and making sure the family comes first. You leave work early for family events.

Pleasure—If you are pleasure driven, your main concern is your personal pleasure: What fun is doing this for me?

Friends—If you are friend driven, your main concern is your friends and your relationships with them. What can we do as a group after work or on weekends?

Work—If you are work driven your main concern is getting the job done right. The job/company comes first; you are a company person.

In addition to the personal principles, there are engineering work principles that guide our decisions during work and greatly affect the quality of our work and our career path. In no particular order, these are:

Technical excellence—If you are driven by a desire to achieve technical excellence, your main concern is doing a superior technical job. The science of the project comes before all else.

Power and status—If you are power and status driven, your main concern is obtaining power, title, awards, and prestige.

Growth and learning—If you are driven by a desire to grow and learn, your main concern is continually increasing your knowledge about new areas of engineering—knowledge for knowledge's sake.

Self—If you are self driven, your main concern is what's in the assignment for you. How do you benefit? Is the reward worth the effort?

Company—If you are company driven, your main concern is the company. The company comes first above all else.

Helping others—If you are driven by a desire to help others, your main concern is making sure everyone gets along, the team above all else. Helping others learn and get ahead is a major concern in your life.

Before you can come up with an action plan, you must determine which principles drive your decisions. The best way to do this is to rank the principles. There is no right or wrong ranking; the order will be determined by you and your beliefs.

MISSION STATEMENT AND GOALS

Once you have identified and ranked your principles, the next step is to generate a mission statement. I recommend that you generate both a personal and a work mission statement. Here is an example of each.

My *personal mission* is to live with integrity and make a difference in the lives of others. To fulfill this personal mission, I will make my family and spouse number one in my life and my work/career number two. I will work hard to balance family and work and will choose relationships with people as most important. I will work hard to provide my family the highest quality of life I can, spiritually, emotionally, and physically. I will lead my family by example.

My *work mission* is to do the best engineering job of which I am capable, to update my skills, and not to settle for inferior work. I will develop my skills and knowledge for the purpose of some day leading an engineering group to develop superior, safe and cost-effective products for my company. To fulfill this work mission, I will apply sound engineering principles and seek ongoing education to update and expand my knowledge of engineering and business. I will work to develop people as well as products, and I will demand the same level of performance from myself as I require from others. I will choose to find the good in things before I find the bad, and I will be honest. I will honor others' performance and place team performance over my personal work goals.

Generation of a mission statement will require several iterations, which may take several weeks. Factors to consider include the balance we wish to attain, the stage of our engineering career, the ultimate position we want to hold, and any danger signs we may perceive.

The experts are often quoted as saying that balance in life is important. They identify the following factors to consider in achieving balance:

1. Religion
2. Family/Spouse
3. Self
4. Work

Maintaining balance in your life is one of the most important, but hardest, things to do. To be a good engineer, you need to maintain balance among family, work, health, social activities, and religion. As shown in Figure 3, your career plans must include a balance between personal life and the demands of the job. Also, the equation for success has two major components, as shown in Figure 4. Total success in life is equal to success in your personal life plus success in your engineering life. You cannot measure success solely on just one or the other. Your real challenge is to determine what will make you successful in both. Then take action to achieve that success.

Another aspect to consider during the generation of your mission statement and subsequent action plan is where you are in your career. Shown in Figure 5 are typical goals that engineers have in their personal and work lives. These examples should help you identify your own goals. Note, however, that goals change as we move through life; therefore, continual reassessment is important.

One very important decision that young professional engineers face in their careers

Career Plans Must Include Balance Between Work & Personal Life

- **Health**
- **Family**
- **Vacation**
- **Hobbies/ Stress Relief**

- **Deadlines**
- **Workload**
- **Overtime**
- **Training**

To Ignore or Deny This Balance is Setting Oneself Up for Disappointment

FIGURE 3. Balance between personal life and work.

There Are Two Parts To The Equation of Success

Success & Happiness **=** Career Technical or Business Success **+** Personal Success

FIGURE 4. Equation to success.

is whether to remain technical or go into management. The actions required for these two career paths are shown in Figure 6. Your career goals and plans will be greatly affected by this decision.

Danger signs are the final area to consider when writing your mission statement and career goals. These danger signs are clear indications that something is out of balance, and it is time for replanning. Here are some danger signs that should not be ignored:

- Your career growth has stopped, or you are receiving poor ratings.
- Excessive hours at work; 50 to 60 hours per week is the norm.
- Excessive fighting at work with coworkers and/or at home with family/spouse.
- Work is no longer fun; conversely, staying at work to avoid going home.
- Economic downturn in industry, downsizing, and layoffs.

If any of these danger signs apply to you, it is time to take corrective action! This is similar to running tests in your engineering work. If you find that the equipment is failing, you immediately do analysis, determine the problem, and rework/redesign the equipment. Your career should not be considered any different in this regard; danger signs are simply signs of failure pointing up the need for adjustment.

After you have identified and prioritized your principles, you need to generate a list of goals based on the principles. The goals should be achievements you plan to accomplish in your career. Following this, you need to generate specific actions for each of your goals. The actions should be well defined and have completion dates associated with them. And, finally, you need to start taking action. You will adjust your goals as your career develops.

Years From College Graduation	0-5 Years	5 - 10 Years	10 - 20 years	20 - 30 Years	30 - 40 Years
	First Job	Early Career	Mid - Career	Late Career	Retirement
Age	22 -30	30 -35	35 - 45	45 - 55	55+
Company / Technical Career Goals	• Adjust to Work Environment • Learn Company Ropes • Enhance Technical & Business • Career Planning	• Focus on Technical or Business Specialty • Develop Technical Skills • Seek out Higher Levels of Responsibility • Publish Papers • Lead Product Development	• Make career path Decision Tech .vs. Business • Develop Leadership Skills • Update Training • Supervisor • Return For MBA • Publish Papers • Patents	• Continue Leadership Development • Update Technical • Upper Management • Mentoring Junior People • Start Consulting Career • Senior Role in Company - Staff	• Leveling of Career and Less Responsibility • Consulting Role • Teach Classes
Personal & Family Goals	• Payoff College Debt • Have Fun • Financial Planning	• Marriage • Purchase Home • Start Family	• Family Vacations • Child Development • Children School & Sport Activities	• Family Vacations • Plan For Retirement • Children College Education	• Plan For Retirement • Wedding of Children • Grand Kids • Less Pressure

FIGURE 5. Examples of typical goals throughout life.

121

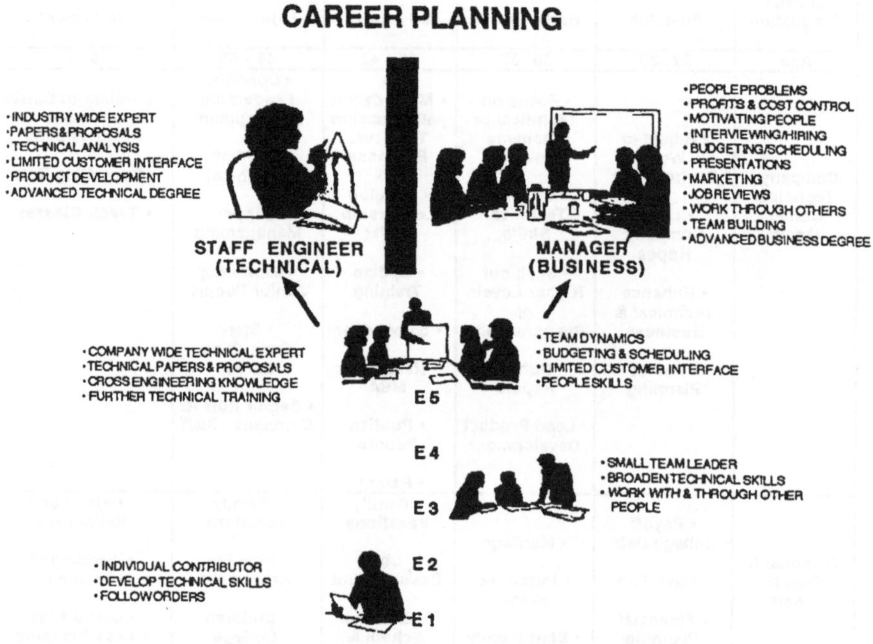

CAREER PLANNING

- INDUSTRY WIDE EXPERT
- PAPERS & PROPOSALS
- TECHNICAL ANALYSIS
- LIMITED CUSTOMER INTERFACE
- PRODUCT DEVELOPMENT
- ADVANCED TECHNICAL DEGREE

- PEOPLE PROBLEMS
- PROFITS & COST CONTROL
- MOTIVATING PEOPLE
- INTERVIEWING/HIRING
- BUDGETING/SCHEDULING
- PRESENTATIONS
- MARKETING
- JOB REVIEWS
- WORK THROUGH OTHERS
- TEAM BUILDING
- ADVANCED BUSINESS DEGREE

STAFF ENGINEER (TECHNICAL)

MANAGER (BUSINESS)

- COMPANY WIDE TECHNICAL EXPERT
- TECHNICAL PAPERS & PROPOSALS
- CROSS ENGINEERING KNOWLEDGE
- FURTHER TECHNICAL TRAINING

- TEAM DYNAMICS
- BUDGETING & SCHEDULING
- LIMITED CUSTOMER INTERFACE
- PEOPLE SKILLS

E 5

E 4

E 3

- SMALL TEAM LEADER
- BROADEN TECHNICAL SKILLS
- WORK WITH & THROUGH OTHER PEOPLE

- INDIVIDUAL CONTRIBUTOR
- DEVELOP TECHNICAL SKILLS
- FOLLOW ORDERS

E 2

E 1

FIGURE 6. Career path (technical vs. business).

Principles	Goals	Actions
Religion	Attend Services	Attend Weekly/Sunday
	Volunteer Help to Poor	Serve Food Shelter 2/year
Family	Weekly Family Activities	Read to Kids 2/week
		Train Kids on Computer
		Make home for Dinner 4/week
		Take kids to sporting events
		Plan Family Vacation
		Visit School
	Financial Security for Family	Savings/Kids Education
Work		
Self	Get Raise	Determine Criteria For Promotion
		Get Feedback On Improvement areas
		Update resume/interview
		Improve work Quality
	Training	Attend Seminar
	Career Goals	Decide Technical vs. Mgmt
		Publish Paper
		Lead Product develop team

FIGURE 7. Example principles/goal/action sheet.

To aid in this process, a form like the one in Figure 7 can be completed. In the left column are the principles and values around which you wish to structure your life. In the next column are goals you hope to achieve during your career. And in the last column are the actions you will take to accomplish these goals.

At this point the process may seem a bit overwhelming. The first cut at your career plan is not nearly as high quality and complete as you thought it would be. Remember, this is a life-long process of continual refinement. You have only taken the first step. The product is not as important as going through the process. Keep working at it continually. The most important benefit is getting organized, identifying what is important to you, and coming up with a plan. I cannot stress enough that failing to plan is simply planning to fail!

ACTION!

After you have completed your strategic career plan, you come to the hard part, taking action! To get started, pick one or two easy actions that you can complete easily.

This is similar to an athlete warming up before the big event. It helps you get started and gives you the momentum to continue.

Finally, are you in the right job? You will quickly realize the answer to this question once you have a strategic career plan.

ABOUT THE AUTHOR

John Hoschette is a senior staff engineer with Lockheed Martin in Sunnyvale, California. Over the past 23 years, his career has covered such areas as developing infrared sensors for night vision systems and laser sensors for weapons and helmet-mounted displays. He holds BSEE and MSEE degrees from the University of Minnesota as well as a Business Administration Certificate. He is a member of the IEEE-USA Career Maintenance and Development Committee and is PACE Chair for the Santa Clara Valley Section.

Mr. Hoschette has written a book, *Career Advancement and Survival for Engineers* (ISBN 0-471-01727-2). He has taught career development seminars at the University of Massachusetts, Lowell, and at Merrimack, Clark, Tufts, and Brown Universities. You may contact him by e-mail at ctsgroup2@aol.com.

Innovate for Today

B. KRAUSE

ABSTRACT

Contrary to popular belief, the greatest innovations that have changed the ways we do things and that have become standards in the industry have not been major breakthrough concepts that have revolutionized society. Societal revolutions take time and become implemented quite slowly. Rather, most innovations have been simple, minor and easy to implement, using resources that are already in place. This presentation will review basic concepts of innovation and will present recent examples of innovative products, services, societal practices, or processes that have occurred that make life easier, more practical, or were just plain good ideas whose time had come. These examples have been gathered through surveys and interviews to find the most common examples that come to mind of the average person on the street.

Examples will also be given of products, services, societal practices, or processes that have been found in dire need of innovation—what doesn't work, doesn't work well, or what ought to be done in a totally different way. What will be shown here will be a list of problems that need solutions, rather than solutions to problems. The main thesis is that innovative creativity from today's engineers for today's problems can bring innovative products, services, societal practices, and processes today, not tomorrow—but only if we innovate for today.

INNOVATION

"Innovation is the specific tool of entrepreneurs, the means by which they exploit change as an opportunity for a different business or a different service. It is capable of being presented as a discipline, capable of being learned, capable of being practiced." (Peter Drucker)

Bob Krause, Past Chair, IEEE-USA Career Maintenance and Development Committee

PRINCIPLE OF INNOVATION

Viewing innovation as a disciple allows us to apply principles that will help us innovate. Purposeful, systematic innovation begins with the analysis of the opportunities. Innovation is both conceptual and perceptual; look, ask, and listen. To be effective, an innovation has to be simple and focused. Effective innovations start small. A successful innovation aims at leadership.

As with any set of principles, there are do's and don'ts. The don'ts are: (a) Try not to be clever; (b) Don't diversify, don't splinter, don't try to do too many things at once; (c) Don't try to innovate for the future; innovate for the present. The three conditions for success, or the do's, are: (1) Innovation is work; (2) To succeed, innovators must build on their strengths; and (3) Innovation is an effect in economy and society, a change in the behavior of customers, of teachers, of farmers, of eye surgeons—of people in general.

To prime the pump, I did a survey to see what came to mind when the average person (engineer and non-engineer) is asked for examples of recent innovations that have changed the way we do things and that have become standards in the industry. As expected, I received some very broad responses that were way off the mark, some that were close, and some that were very insightful. I then asked for examples where innovation is direly needed—things that don't work the way they should or that should be done in a totally different way. The results follow.

RECENT INNOVATIONS

- juice cartons with re-cappable closures
- sticky notes
- aerosol cans to spray away dust
- silk flowers for those who don't have a green thumb
- programmable thermostats
- support groups
- telephone redial button
- offering step-by-step list of driving instructions at kiosks of Hertz rental cars, for navigating in unfamiliar areas
- offering automated in-vehicle "navigators" (never-lost for Hertz) in rental cars, for about $7.00 extra per day
- disposable pull-up pants for toddlers and seniors
- universal remote control for VCR, TV, etc.
- portable carbon water filters
- portable air pumps (pocket-size)
- adhesive nose patches
- caller ID
- ATMs

- salad in a bag
- oil change in a parking lot (mobile service)
- disposable diapers/razors/cameras
- ziplock storage bags
- hospice program
- home health for seniors
- self-stick stamps
- baby monitors
- recyclables
- cellular phones
- remote door opener for auto
- Nicorette gum for smokers
- super glue
- safety pin paper clips
- v-shaped paper clips
- bar codes word processor
- wheels on suitcases
- wireless modems for laptops
- good heat-and-serve take out meals at grocery stores
- five-day movie rentals
- credit card-dispensed gasoline at pump (RF ID cards)
- ordering hotels, planes, cars via the internet
- mobile phones
- multi-line conference calls
- anti-bacterial hand cleaner
- global world wide web and browsers
- cordless phones, fax
- CNN network
- 401k plans
- contact lenses
- Dockers refrigerator magnets
- plastic containers
- battery backup for clock radios
- cup holders in cars
- residential centers for the elderly
- Velcro magnetic child proof cupboard locks
- combo fax, scanner, printer, copier
- internet computer and accessories
- video transistor satellite
- internal combustion engine
- water collection and redistribution
- cars designed for 100,000 miles without maintenance

- LCD monitor screens
- computer-TV merge mini-satellite dishes
- popcorn that pops in the bag
- model parts for an epoxy soup (STL)
- LEDs bright enough to make traffic signals
- e-mail
- plastic garbage bags
- microwave oatmeal

NEEDING INNOVATION

- a compact fold-up cover to protect your car in a hailstorm
- a hail-proof umbrella
- access to better route-specific traffic reports
- resealable dog food cans
- long-term health care in the home
- daycare in the workplace
- local transit systems
- grocery shopping via the internet or computer with home delivery
- airline service
- nationwide cellular service
- international currency
- road surfaces that last—no potholes after one year
- software that works when you load it
- software without a 3-inch volume of instructions
- airflow system for homes with equal distribution
- replacement for aluminum cans
- voice to computer conversion that works
- replacement for blue jeans
- loading and unloading of airplanes
- replacement for carbonated drinks
- inexpensive universal access to telephone networks
- limitations on liability (excessive lawsuits)
- better educational processes in elementary and secondary schools
- secure web commerce
- restrain infotainment element in journalism/media
- develop culture based on individual worth
- voice message systems
- job searching—need a better way to bring skills/jobs together
- low water use toilets
- sailboats—light, safe, and easy to sail
- public surface transportation—improve efficiency and appeal
- Internet—speed up access to information

- federal tax rules
- banking
- better communications antenna systems
- better storage battery or stationary electrical source
- highway systems
- weather forecasting
- city traffic control systems
- interior lighting
- U.S. railroad system
- individual brainpower utilization
- world patents
- lawn furniture
- dust and pollen filters
- pest control

The results showed that we still tend to think of innovation as a grander concept that happens in a broader spectrum, something that will be a breakthrough concept that revolutionizes the world. But innovation of that nature is really a collection of separate innovations and cultural and social upheavals that allowed those separate innovations to be adapted. The list of items needing innovation followed the same pattern, showing our mindset of seeing problems in a grander scale. However, it did show our recognition that innovation is needed in some of the every day aspects of our lives.

The innovations that are the most successful are those that satisfy a need today and result in changes that can be adopted today, such as the paper clip or ziplock bags. As engineers we should innovate for today.

ABOUT THE AUTHOR

Bob Krause is a registered professional engineer with a bachelor's degree in electrical engineering and a master's degree in business administration. He has more than 20 years of experience in analysis of accounting, economic, rate making, regulatory, and contractual matters. He has testified as an expert witness before state and federal agencies and has served on task forces in two states that have sought solutions to industry problems. He has consulted internationally, performing privatization of power studies for the government of Pakistan and financial analysis for the government of Egypt through the United States Agency for International Development (USAID) and the World Bank. He is skilled in analysis, management, planning, contracts, seminars and training, and problem solving.

Mr. Krause is active in IEEE, serving as the past chair of the IEEE-USA Career Maintenance and Development Committee. He founded the first IEEE Consultants' Network in Texas and is on the executive committee of the Alliance of IEEE Consultants' Networks.

Yes, "Tekkies" Can Talk— and Sometimes Even Sell

T. LEECH

ABSTRACT

For many, the standard expectation of presentations made by engineers and other technical professionals is low. The common stereotype is that the presentation will be boring and that the speaker will be bright, nerdy and not likely to be a snappy presenter. A common lament of sales managers is "I hate to take an engineer to a marketing meeting." Surveys of top managers have shown that communication skills for engineers are highly important, yet capabilities are weak.

However, by necessity, many engineers (for this article, I've grouped all "tekkies" as engineers) have developed their abilities to present effectively and to sell their ideas and programs in the process. So what specific techniques separate the drab engineers from the persuasive? Here are some tips, provided by an engineer who has made a career out of presenting and selling and helping others improve and apply those skills.

ENGINEERS AND PERSUASIVE COMMUNICATION— AN UNLIKELY MATCH?

Are Engineers Good Presenters?

As reported in *Electronic Engineering Times* (November 7, 1983), an in-depth study by AT&T Corporation found that engineers were deficient in key skills needed for management. Paramount among those skills were "neglected non-technical skills—such as planning and organizing, written and oral communications, sensitivity and persuasiveness."

In another survey noted in *Engineering Education* (November 1979), executives ranked the 10 most important skills engineers (EE & CE) should have mastered when they graduated from college. At the *top* of all three disciplines was ability to communi-

Thomas Leech, Principal, Thomas Leech & Associates

cate. Then the executives rated the capabilities of their engineering new hires for each of the 10 skills. At the *bottom* was that same category—ability to communicate.

These studies are often anecdotally supported in conversations with executives, who readily describe engineers or managers who are technically proficient but who are weak when it comes to being able to present their ideas and knowledge. This weakness may limit value to the organization and their own opportunities. When it comes time for promotion or more challenging assignments, this weakness may keep the poor presenter from being chosen for such opportunities.

Many Engineers Have Become Good Presenters

It doesn't take long on the job for most engineers to realize they need to polish their presentation skills. Their job often entails presentations, including giving status reports, participating in program reviews, selling ideas, leading a team, and training others. One application that can help engineers hone their presentation skills and that can enhance their careers while serving a vital need of their organization is working on proposal teams. Personnel assigned to key management or technical slots on these teams are selected largely because of their communications ability, as these positions often require the team members to make oral presentations during the selection process. All team members have to be able to sell the selection board on their capability before they're able to exercise it on a contract.

In many successful organizations, engineers play major roles and must apply presentations and selling skills on a regular basis. Job slots such as chief engineer, program director, applications engineering manager, new business development specialist, sales engineer, CEO and entrepreneur require people with technical, management and presentation capabilities. So apparently, "tekkies" can talk *and* sell.

TEN TIPS FOR BECOMING A PERSUASIVE PRESENTER

1. Determine what you want the presentation to achieve, and ask for it.

Being able to state the presentation objective clearly does not seem to be much to expect. However, the inability to do this is a source of frequent complaint by the top-level executives who make up presentation audiences. Paraphrasing one vice president, "They tend to show up with charts in hand and unload lots of data, but without an apparent end goal. I ask them and their answer is often a stammered 'Well, isn't it obvious?' "

In our training programs, we ask participants to state a presentation objective clearly. Initially, they often start out with "To win a contract." Upon further analysis, they learn that this is almost never the actual outcome of such presentations, making their objective unrealistic. When developing a presentation objective, remember these two keys:

- Define the objective in terms of what you want the audience to do if they accept your proposition (assuming this is a selling presentation); and
- Make it truly achievable—and measurable, if possible.

So in this case, perhaps a better goal would be "The key customer will commit a $10,000 contract add-on for a specific trade study."

Another common lament from executives about technical presenters is the presenters' apparent timidity when it comes to stating explicitly the logical conclusions and specific actions they want the audience to take. So, as the dreaded "sales world" adage goes, "Ask for the order."

2. Develop an audience-based strategy.

Engineers' presentations are typically very strong in covering the process. We feel both the obligation and the inclination to explain in detail how we went about something: what clever analytical methods were used, what trade studies were performed, how the project was organized and executed, and all the wonderful features of the resulting end product. The main difficulty with this approach is that many in the audience—especially those who have advanced to higher positions in the organization and who are now the principal listeners and decision makers—don't have the time to listen to all this detail; don't find the in-depth discussions provide the key information they need; and may not be able to grasp all the technical depth anymore anyway. This might explain why so many of them tell me that many engineers spend too much time talking about the wrong information.

But if the information we cover typically is "wrong," what is the missing and right information? The key is contained in the old adage that stands paramount in the minds of almost any audience member: "What will it do for me?" To help answer this question, be sure your presentation:

- Stresses critical vs. inconsequential issues;
- Focuses more on "why" and less on "how"; and
- Emphasizes benefits and applications (audience-focused) vs. features (speaker-focused information).

3. Organize material so top management can absorb it readily.

This concept was put to me a few decades back from my own general manager, and I've heard it reinforced on other occasions by other high-level executives. "Engineers in particular so often present material as if it were a mystery novel, requiring us to sit through 250 pages (or 25 charts) to find out the butler did it." What he suggested as a much better approach was to present in the manner of a newspaper reporter, who immediately lets the reader know the butler did it.

Consider how presentations are organized. The mystery novel approach equates to the format followed in many standard engineering reports: lay out the premise, state the assumptions, cover the literature search, analyze the various options, and finally provide the results. The journalistic approach starts with the results, skips lightly across the heavy details (such as assumptions and literature search), and then hits the essential topics to make the case. This latter approach would accommodate the specific request made by many time-pressed upper management executives for a *summary* of the presentation.

An executive vice president once said, "I always take homework to presentations by engineers. As they ramble on about information of no value to me, I read articles, review outgoing letters, or catch up on my mail. When they finally arrive at the information I need, I start to listen."

4 It's not all data that sells; weave in stories, examples and scenarios.

One of the major culprits in weak technical presentations is the heavy emphasis on hard, quantitative data and the plethora of line graphs and other data displays. These may be important information items. However, other forms can be equally as valuable—or more valuable—for providing clarity, interest, argument and motivation. These include vivid relevant examples, anecdotes about individuals or organizational successes or disasters; and hypothetical illustrations or scenarios. Yet these are often the tools engineers are reluctant to incorporate. It's more technically comfortable to rely on a table of statistics about product safety deficiencies than relate a story about a tragic accident that resulted from the product. The most persuasive speakers use various anecdotal and illustrative tools well.

5. Use visual aids to sell, not obfuscate.

A colleague and I were part of a large audience for a presentation made by the head of a large, multi-national corporation. It was computer-based and projected onto several large screens. About midway into the development of a key point, he displayed a visual on screen and asked "Now, what's wrong with this chart?" My associate and I instantly whispered to each other, "What's wrong is, you can't read it." That was not the lesson he intended us to draw. Listen to almost any executive discussing the last presentation given by an engineer and he or she will generally note that the speaker included several barely readable charts—"eye charts." Even with the best computer software, output devices, fonts and colors, the fundamental rule of readability is often violated.

Right behind the readability problem is the seeming need to display the whole story. If anything, engineers are thorough and detail-oriented. Detail unfortunately often obscures key information. The old KISS (Keep It Simple, Sam) formula can be applied here to good advantage, especially when combined with the FOCUS (focus) principle.

When asked to state what conclusion the listener should draw from the multi-entry table or graph displayed on the screen, some technical presenters are prone to comment "The data speaks for itself." That may be the case, but the data may send different messages to different people. Therefore, to preclude the receiver from possible misinterpretation, clarifying summary titles are often valuable.

6. Instill an image of professionalism by your readiness to present.

Who has not experienced a delay in the start of a presentation because the speaker has just discovered the projector won't work because the bulb is burned out or because there is no projector? As audience members, we may find these situations amusing (especially when our competition is having the problem), frustrating (as the clock continues to run with nothing happening), or even decisive (as a vivid demonstration about how well this team will perform current or proposed work).

"The devil is in the details" has been cited as a valuable credo. "And he or she goes by the name of Murphy," can be applied to presentations. Marshall McLuhan's "The medium is the message" made an important point a few decades back. (We remember the burnt out bulb long after the technical content.) There is truth to all of these statements, so the astute presenter must make "check and recheck" standard operation. In this way, the initial, continuing and lasting impressions will be positive, not negative.

A key part of preparation that is often skipped is practice. I recently was speaker coach to a team pursuing a significant defense contract. An oral presentation was a required and important part of the proposal effort. Having worked on a few hundred of these, I've long realized the chances of success without rigorous rehearsals is low. Yet, getting the team to practice was not easy. We finally got the team to do their first practice, with a predictably mediocre performance. Still they dragged their heels for the second, which was still deficient. Their final practice came out well. Following the actual presentation, the team head said, "You know, I was reluctant to do those dry runs. We were busy and I didn't think they were necessary. But doing them proved they were absolutely necessary. We'd have been killed without those practice sessions."

Many technical presenters dismiss the need for rehearsals, with such comments as "I don't need it. I know this material well. I don't have time anyway. It is a waste of time." Unfortunately, without practice, their presentations will suffer.

7. Remember the "how" of presentation can be as important as the "what."

To many technical presenters, concern with delivery is equated with flim-flam or used-car salesman techniques. Speakers prone to delivery "theatrics" or "dressing for success" are often disdained.

A concept I first heard credited to poet Emily Dickinson sums up the essence of why engineers are often drab presenters. She cited a fellow poet who "had the facts, but not the phosphorescence." For many engineers, comfort lies in displaying reams of information while providing a spoken commentary notable for its drab, unemotional delivery; big on facts, tiny on phosphorescence.

A common segment used on television shows or commercials to amuse viewers shows a lecturer whose monotonic manner has lulled the audience into a near slumber state. Unfortunately, this delivery style is what many engineers apply in the conference room. Often this is part of a deadly triple whammy—monotone, busy slides read verbatim, and talking to the screen instead of audience. These are guaranteed to do in even the most well-intentioned receivers. Throw in a darkened room at 1:00 p.m. and it's "zzzzzzz" time.

Well-known communication studies show the high degree of importance listeners place on spoken vs. nonspoken presentation. In classes and seminars, when I ask students—many of whom are engineers—to rank these factors, they generally undervalue non-spoken factors, which include tone of voice, inflection and body language, and which all add up to "phosphorescence."

8. Prepare for the informal as well as the formal.

The outcome of many presentations rests as much on how well the speaker succeeds during the audience grilling called Q&A (question & answer) as with the fully prepared and polished formal presentation. "No question, we won it on the Q&A," said one CEO. An executive vice president had the opposite experience: "No question, we blew it in the Q&A. We could see the contract slipping away on one question that we were not able to answer satisfactorily." This is another area in which technical presenters often come up short. Almost any program manager or executive can offer anecdotes about the engineer who messed up during Q&A, with a common example being a person asking for the time and the engineer launching off about how to make a watch. This goes back to Tip #2: focusing too much on features (how) and not enough on benefits (why).

Good presenters will prepare as hard for the Q&A as for the presentation itself. After all, President Clinton and his advisory team do this before critical press conferences, and Bill Gates did it before his recent much-publicized Congressional hearing. So doesn't it make sense for us other tekkies to make this part of our preparation discipline? Here are some preparation tips:

- Think about what sorts of questions might come up;
- Prepare for those potential questions, perhaps even developing additional backup charts;
- Practice answering those questions—both the response (what) and the process (how).

9. Get smarter; fill in the educational gaps.

The first eight tips are specific to preparing and delivering presentations themselves. Tekkies are especially good at adding to their technical, professional knowledge base. Becoming a well-rounded and proficient practitioner of presentations and persuasion means adding to that knowledge base. Here are some ways to do that:

- Join Toastmasters;
- Take a college course or training seminar in public speaking/presentations;
- Take a course in marketing;
- Add some presentations, marketing and motivation books to your library (and read them). Here are some starters: my book on presentations (noted in the About the Author section); *Strategic Selling* by Miller & Heiman; *Positioning To Win* by Beveridge & Velton; *Hierarchy of Needs* by Abraham Maslow.

10. Make "become a skilled presenter" part of your career growth plan.

Finally, a long-term plan that incorporates colleagues and management can provide the necessary on-going foundation for implementing all the above suggestions:

- Enlist management support for performance reviews;

- Become a student of all presentations (yours and others'), to pick up continuing lessons-learned;
- Seek out presentation opportunities (professional papers, teaching, speakers' bureau, community causes, church, etc.);
- Request feedback from audiences, including management and associates;
- Become an internal presentations "coach" for colleagues.

SUMMARY

Yes, Tekkies Can Talk—and Must, in Order to Succeed

Any organization with technical products and services must, by necessity, have strong presentation capabilities among personnel in many disciplines and specialties. Many professionals in technical fields have realized the importance of presentations to their value and advancement. Because of this, they have enhanced their technical proficiency with presentations expertise. The 10 tips noted here can provide a basis for any "tekkie" to get moving toward presentation success. Do you have any reason not to start right away?

ABOUT THE AUTHOR

Thomas Leech is the author of *How to Prepare, Stage & Deliver Winning Presentations* (AMACOM—American Management Association), 2nd edition, 1993. He has headed his own firm since 1980, providing presentations consulting, training seminars and conference programs for organizations nationwide, with a particular emphasis on aiding technical firms and teams. For 14 years he has taught team presentation skills for the University of California at San Diego Executive Program for Scientists & Engineers, and teaches presentations for the Project Manager's Program at the University of San Diego.

Mr. Leech worked previously for 20 years at General Dynamics, with assignments in business development, communications and engineering. He has a bachelor's degree in aeronautical engineering from Purdue University and a master's in management science from United States International University.

For further information or to discuss the specifics of this paper, contact the author at Thomas Leech & Associates, 4901 Morena Blvd. #102A, San Diego, CA 92117; telephone (619) 274-5668, fax (619) 274-5669, e-mail winpres@aol.com.

People Skills in a Competitive Environment

J. V. LILLIE

ABSTRACT

This paper will present information that should be useful to the professional development of individuals who work in technical fields. The competitive environment will be defined, people skills will be identified, and ways to apply these skills will be discussed. Methods of skills assessment will be presented, including the idea of using performance evaluations to improve skills. A skills self-assessment will be discussed, along with ideas for skills improvement.

THE COMPETITIVE ENVIRONMENT

The term "competitive environment" is used to define the workplace as it exists today. Over the past few years I have had the privilege of traveling around the United States on business with BellSouth, as a volunteer for the Institute of Electrical and Electronics Engineers (IEEE), and, at times, on vacation. During this travel I have been able to meet with many people in many different places—people with a wide variety of engineering jobs. One topic that seems to be on everyone's mind is job security. As I have talked with these people I have found that they are concerned not so much with employment but with the change that comes with or is caused by employment.

Customers require that companies deliver products and services at competitive prices. To accomplish this, companies must evaluate all components that add cost to the product or service they offer. One of these costs is the cost of people; at service companies, the cost of people can easily exceed all other costs combined.

While being competitive, companies create a competitive environment for their employees. They do this through such activities as downsizing, merging, re-engineering, right-sizing, and outsourcing. There may be other names for these processes and activities, but the bottom line is that companies are reducing staff to reduce the overall cost of the product or service they provide.

Joseph V. Lillie, Area Manager, BellSouth Telecommunications, and IEEE-USA Vice President, Professional Activities

When companies reduce staff by downsizing, the employees who are retained must compete for the positions that remain. When companies merge, employees in similar positions must compete for the merged positions. When companies re-engineer, they change processes, which usually reduces their need for employees. When companies right-size, they eliminate the excess of employees that has developed over time. When companies outsource, they move work from an employee group to an independent vendor.

Each time one of these changes occurs, the impacted employees must compete for the remaining positions. Since we do not always receive advance warning about employee reduction plans, we must always be prepared for such an event. We must always assume that we are working in a competitive environment.

The actual selection process used to determine the surviving employee group will vary with the situation and the company. Regardless of the specific process used, evaluating employees will include skills assessment. Included in this assessment will be a review of each individual's technical knowledge, including educational background and demonstrated ability. The evaluation will also include an assessment of the non-technical skills possessed by the employee. These are the skills that we use to apply our technical knowledge; they are "people skills." The people-skills evaluation could well drive the decision of who stays and who goes.

SKILLS IDENTIFICATION

People skills are the non-technical behaviors we use to promote our technical usefulness. If we were to use a brainstorming process, we could develop a list of skills that would fit into the category of people skills. Following are descriptions of people skills that I consider important. This is not an exhaustive list, but rather a representation of the skills used in my daily work environment. These skills include communications, teamwork, negotiating, manners, ethics, attitude, and humor.

COMMUNICATIONS

Each of us needs to communicate with others on a regular basis. Such communication may be written or verbal and may be conducted in a formal or an informal setting.

Informal written communications include project notes, e-mail, and office memos. Formal written communications include letters, proposals, project reports, and professional papers. Informal oral communications include individual discussions, group discussions, and day-to-day conversations, while formal oral communications include project presentations and public speaking events.

If you do not possess and use good communication skills and techniques, you may not be able to sell a project you developed. It becomes difficult to implement a technical solution to a complex problem when you do not have the skills necessary to convince others that the solution is workable. A great technical discovery may remain on the shelf if the right message is not delivered when the project is presented and discussed.

You can learn effective communication skills by attending training sessions, but you must apply them to develop them. You can and should participate in non-work settings to develop and apply these skills. For example, you can volunteer to serve on boards of non-profit organizations, get involved in your church, or participate in activities at your children's schools. In each of these volunteer activities, look for opportunities to use communication skills by making presentations, writing letters, and leading group discussions. In doing this, you will improve your communication skills. It also helps to develop relationships with others in these organizations, so you can receive constructive feedback. In this way, you can learn from mistakes you make in a non-work environment and then avoid making the same mistakes in a work setting.

You need to be prepared for the day the company leaders are visiting your office and the person scheduled to make a presentation calls in sick. By developing confidence in your communication skills, you can step in and handle the task. These and similar opportunities can occur with little advance notice; if you are willing to perform in such situations, you will be a great asset to your employer.

TEAMWORK

Teamwork involves working in groups to solve problems. None of us possesses all the technical knowledge required to complete complex tasks. To overcome shortfalls, we lean on others who posses the required knowledge, and others lean on us to supplement their shortage. The synergy of this interaction allows us to accomplish more as a team then we could accomplish as individuals.

To work effectively in teams we must be willing to listen to and respect others' ideas and views. We must be willing to contribute fully to the process and work toward the goals and commitments established by the team. When we work as a team we can accomplish such tasks as sending astronauts to the moon and returning them safely.

We need to apply the group approach that we used so well as students. I know that, when I was a student, I almost always worked on homework assignments as part of a team. This process worked very well; now as employees, we need to continue to make this process work well.

Companies tend to look for individual accomplishment when determining salary increases. We must embrace the team concept and be willing to talk about group accomplishments even when our supervisors are discussing salary increases. We need to get rid of the "I" and start using "we" when we talk about accomplishments. We need to be part of the team.

NEGOTIATING

Any time people are placed in a situation of having to work together, it is possible that they will have differences. We need to accept the fact that there is usually more than one right way to do things. When involved in a difference of opinion we should analyze the situation and determine what we are willing to give up and what we will not sacrifice. As we negotiate, we need to look for a win-win situation.

When negotiating we should always leave an "out" for the other person. It is important that we recognize the fact that this will not be the last time we negotiate with this person. Therefore, we need to avoid backing them into a corner, where their only option is to attack; if they attack, we may lose. If we give them an out, we can both win.

MANNERS

It is important to have good manners and be polite when in public places, including the office. You should have learned good manners at home from your parents. If you didn't, then it is important to make this a priority. Good manners need to become automatic, because you never know when they may pay dividends in the future.

During your professional life you will need to prepare for and attend many social events. It is possible that your boss' spouse will attend some of these functions. When you exhibit poor manners in such settings, you risk becoming the topic of conversation during the drive home. It would be terrible if you missed out on a great job assignment or a promotion because the boss' husband or wife pointed out that you exhibited poor manners. Don't kid yourself; it can and does happen.

ETHICS

As engineers, we must conduct ourselves in an ethical manner at all times. When discussing ethics, the question of what ethics is usually comes up. To me, the answer is somewhat simple—ethics is doing the right thing every time. I have done some research on this subject, and I agree with those who say that ethics has its basis in such sacred books and writings as the Bible, the Ten Commandments, the Mishanh Toran, and the Koran. It is my belief that ethics comes from God. If you and I do not believe in the same god that is all right; let ethics come from your god. In either case, someone much more powerful than us knows the difference between right and wrong.

One item headlining the news frequently these days is sexual harassment. To me this is an ethical issue. If you are guilty of sexual harassment, then you have not done what is right and you have not been ethical. The specific activity is not important; the fact is, if you think an act may be unethical, it probably is.

ATTITUDE

You may not think of attitude as a skill, but I feel that it is. I also feel that each of us possesses the ability to select our attitude each morning. When selecting our attitude for the day, we only have two choices—good and bad. With only two choices, making the decision is easy. However, the consequences of making the wrong selection can be devastating.

Attitude is contagious. Your good attitude will spread, as will your bad attitude. Given the choice of spreading good or bad, I think that we would all elect to spread the good. If this is true, then why does anyone show up at work in the morning with a bad attitude? The answer is simple: they decided to.

I have clear memories of going to work with the wrong attitude. Several years ago things were not going my way at work, and I was always willing to share my feelings with others. I started to notice that others were agreeing with me. The problem was that none of us was willing to do anything constructive to fix the real problem. *We* were the problem.

I also have clear memories of what happened when I decided that I was going to be happy at work. I decided to have a good attitude. In doing so, I became more productive. I completed tasks that I didn't seem to have time to complete when I had a bad attitude. I did the job that I was hired to do. Others seemed to do the same.

Attitude is a skill. You select your attitude each morning. And your choice will be contagious. Therefore, I recommend that you decide to have a good attitude tomorrow morning.

HUMOR

Humor is defined as something that is designed to be comical or amusing. Humor can be joke telling, but it may take other forms as well. It may be adding a funny comment during a conversation. It has a place in the office and in business meetings. Used at the correct time, a humorous comment can reduce stress and help people relax. When humor is used properly, the stage can be set for detailed discussions on serious matters.

Humor can also be jokes told at appropriate times. It may be appropriate to tell a series of jokes in the office break room or during business meals, but be sure you evaluate the effect your jokes may have on others before you tell them. Others may consider your joke to be harassment. Select your jokes cautiously and use them in the correct setting.

I feel that humor is required in the workplace and that a humorous person can be an asset to an organization. The skill aspect of humor is the proper use of your ability to make others laugh.

SKILLS ASSESSMENT

Communications, teamwork, negotiations, manners, ethics, attitude, and humor are skills that can be the keys to your future. You must accept responsibility for developing your people skills and take steps to ensure that you are ready for the competitive environment.

We can't wait until we are terminated to accept the fact that we are working in a competitive environment and need people skills to survive. So what can you do to improve your people skills? The first step is to perform a self-inventory of your skills. Start with the seven skills outlined here. Be honest with yourself; ask your coworkers, colleagues and members of the group you supervise to provide you with feedback. Ask your family and friends for input; in all likelihood, you behave the same at home as you do at work.

Seek input from your supervisor as well. If you work in an environment where your performance is evaluated and formal feedback is provided on an annual or semi-annual basis, you can use your performance evaluations to gain information about your people skills.

After gathering data on your current ability to apply these skills, develop an improvement plan. This plan should identify specific programs that could be used to improve specific skills. Such programs may be available within your company, and you can attend seminars geared to the skills you have identified as needing improvement. Use all available resources to work through your improvement plan. Reassess your skills performance and continue the improvement process. Skills improvement never ends.

SUMMARY

In summary, consider the following relative to your future:

1. Accept the fact that you are working in a competitive environment.
2. Determine the people skills that are important to your future success.
3. Inventory your current application of these skills.
4. Develop a plan to improve your skills application.
5. Implement your plan to improve your skills.
6. Re-evaluate your performance on a regular basis.

If you are willing to perform these steps, you should improve your chances for success and you may contribute to the success of people who depend on you. Good luck to you in your quest to become a better communicator, team player, and negotiator. Good luck to you as you improve your manners, ethical approach, attitude, and humor.

ABOUT THE AUTHOR

Joseph V. Lillie received a bachelor's degree in electrical engineering from the University of Southwestern Louisiana (USL) in Lafayette, Louisiana in 1974 and a master's degree in telecommunications from USL in 1997. He has worked for BellSouth Telecommunications in Lafayette since 1973. He has held positions of outside plant design engineer, outside plant planner, district support manager, planning manager, engineering manager, engineering/construction manager and area manager. He has attended numerous training sessions on telephony, management, leadership and contract administration.

Mr. Lillie joined IEEE as a student member in 1972. Over the years, he has held leadership positions at the section, region, and national levels. He currently serves on the IEEE-USA Board of Directors as the Vice President for Professional Activities, and he is co-chair of the 1998 IEEE-USA Professional Activities Conference.

Mr. Lillie is a member of Phi Kappa Phi. He has served the Lafayette, Louisiana community in leadership positions with several local organizations. In 1989, Mr. Lillie was named the International Cajun Joke Telling Champion. He and his wife Debbie have been married for 26 years and have two children, Joe II and Jacie, both electrical engineering students at USL.

The Young Professional as Manager: Managing Older Subordinates

S. LOCKHEAD

ABSTRACT

Opportunities will always exist for technical professionals to move into supervisory and management positions. The technology advances of the past few years have allowed an increasing number of Young Professionals (YPs) to assume such roles in their companies. In many cases, recent graduates and young professionals are using cutting-edge tools and methods that provide them with the "advantage" necessary to advance their careers. As new career opportunities continue to arise, the natural progression leads to management positions.

Most college programs concentrate on developing the most technically capable people. Although students can take classes that focus on the business aspects of a technical discipline, these classes tend to only scratch the surface of what young professionals need as managers. Putting the lessons of politics and correctness aside, students often lack many necessary skills. Effective managers must possess many different skill sets; one of the most pressing is the ability to manage subordinates who are older and who have more "years in the business."

Some of the core aspects of the management role will not change. Interpersonal and intraprofessional dynamics will occur when traditional roles are juxtaposed; that is, when a manager is younger than his or her subordinate is. Young professional managers must be attuned to these circumstances and must learn to manage subordinates both as resources as well as individuals. Their success will depend largely upon their ability to "manage" their subordinates and themselves.

INTRODUCTION

The diverse challenges of management can be overwhelming when a YP first steps into the role of supervisor. As engineers, we are all managers of resources. The phrase "What I learned in school was a methodology for describing the application, evaluating

Sean Lockhead, Product Support Group Manager, Kaman Industrial Technologies

possible solutions, testing, formulating, and implementing the perceived best option" is at the heart of the definition of being an engineer. This rationale allows engineers to think about technical problems of any nature. However, as they move into management and supervisory roles, this methodology is no longer clear cut or easily implemented. The changing business climate dictates that engineers know more than just their area of expertise; in essence, they must become a part of the business.

There are opportunities in this landscape for YP engineers to step into a management role—more specifically, supervisor. The skill sets necessary to supervise people effectively have been described by hundreds of people from just about every angle. From Dilbert to Drucker, many have made contributions. However, an area that is becoming more relevant quickly is supervising subordinates who are older. This discussion relates to chronological age, not a perceived age group. As a YP, the term "older" has a wide range of possibilities. The dynamics created by supervising people with more experience and more "years in the business" provide for an interesting journey up the learning curve for any YP.

ESTABLISHING YOURSELF

The first priority of any new supervisor or manager is to establish the command hierarchy. Clearly defined supervisor/subordinate roles alleviate any possible confusion about who has the ultimate ruling in matters. However, the method used to establish these roles is extremely important. Your initial contact and the perceptions you create dictate how others will respond to you as a manager and as an individual. Your management style also plays into the success of your group. Figure 1, taken from *Principle Centered Leadership,* shows paths that result from how you choose to manage.

The path leading to the most successful outcome is the "Principle-Centered" route, which results in a sustained, proactive influence. As author and business consultant Stephen Covey puts it, "The more a leader is honored, respected, and genuinely regarded by others, the more legitimate power he will have with others." The main skills necessary for maneuvering through the managerial minefield successfully include leadership, integrity, team building and maintenance. It is also essential to optimize such processes as empowering and mentoring and to create your group's service level. Every group has a function to perform, whether it is a design, administrative, customer-related, technical support, or other role. A group's ability to perform its function establishes its service level, and the ultimate goal for all groups is to exceed their commitments.

Businesses and work groups often focus on the concept of team-building skills. However, the building process is only half of the puzzle. Teams are built at any snapshot in time. But things change and people change. Groups that sustain a team approach over time will succeed. Their ultimate goal should be to transform their group into a team and then foster the culture within to maintain that mindset.

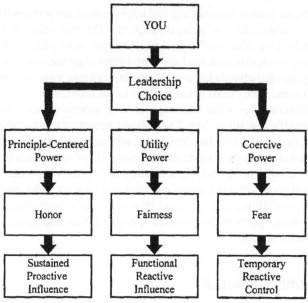

FIGURE 1. Power process.

ESTABLISHING SERVICE LEVELS

Many articles, books, and videos have been produced about customer service and related topics. Using this information, a YP can espouse three precepts to develop leadership ability.

First, *put your primary focus on your people.* Each person needs to feel that he or she is a contributor. People want to develop professionally and personally, regardless of the title they hold. It is your responsibility to create a culture and an atmosphere that offers such opportunity.

Second, *the organization needs to be acceptable to cross-functional and interrelated internal relationships.* Service levels need to center on quality, not quantity. Corporate culture often dictates the extent to which this is possible.

Third, *recognize that the customer relationship draws on the first two areas.* Many say that the customer comes first, that the customer is never wrong, or that customer issues need to be resolved first. By concentrating on your people and your organization, you can reduce customer problems and issues. With the right people and the right mindset, customer-employee interactions tend to be successful and result in satisfied participants.

The YP needs to work to create this approach to service and interaction level. This is best described in the motto of a major hotel company: "Our employees will not treat customers better than we treat our employees."

EMPOWERMENT

One of the most difficult tasks for a YP manager to master is the ability to empower the group effectively. In terms of a YP supervisory role, the effective use of empowerment is better quantified by following a few criteria:

1. *Specify Desired Results*—No empowerment efforts can succeed without first defining what the desired results are. Having a destination or at least a direction is the primary step in empowerment. For example, consider the manager who says, "You're now empowered." Employees would most often respond with, "To do what?"

2. *Set Guidelines*—Give the people the "rules of the game." Include your definition of your supervisory position and authority role. Clear guidelines help to prevent the "I didn't know" excuse.

3. *Identify Resources*—Establish tools, methods, informational libraries, computer technologies, and other resources that may be available to the group. This is an area in which the group can be encouraged to "think outside the box."

4. *Define Accountability*—Empowerment has limits projected through accountability of the results of empowered actions. If everyone were free to act in any way they saw fit, pandemonium would rule. One possible workplace outcome includes a person or group instituting a policy that benefits their area specifically and results in a two-hour time savings in their process. However, is the policy really beneficial if it causes a six-hour time loss somewhere further in the process? Accountability helps facilitate the team concept.

5. *Determining the Consequences*—Feedback closes the loop, allowing for behavior adjustments to be made or events to be analyzed. Again, whole processes need to be evaluated so that the actions of one person do not affect the performance of others adversely. One of the most widely used tools for this type of feedback is a regular review, or as some might put it, "the dreaded yearly review." When used properly, this tool provides feedback and establishes expectations.

OUT OF THE CUBICLE THINKING

The YP must establish an individual style of management that takes into account all surrounding influences. Figure 2 shows concepts that the YP should embrace while making the transition into management and supervisory positions.

Preconceived notions and the inability to maintain an open mind will hinder a YP's transition into management. The on-going balancing act sets limits and introduces guidelines while allowing for flexibility and creativity. However, YPs needs to be cognizant of the fact that this is a gray area.

Some managers love the phrase, "The more outrageous the better" when dealing with ideas. This relates well to the amount of creativity that people generate. The "re-

YP Mindset	
DO NOT	**DO**
* Arrive with preconceived notions.	* Maintain an open mind to all possibilities, no matter how outlandish they might first appear.
* Set restrictive limits.	* Set clear guidelines.
* Fall into the trap of believing that something has to be done a certain way because it has always been done that way.	* Remember that you can use the best of all worlds, including "the old ways" in your approach.
* Stifle creativity by creating rigid boundaries.	* Use imagination and innovation when attacking applications, problems, or issues.
* Assume fear of new technologies will be present. This is a dangerous supposition.	* Show how the integration of technology will increase productivity.
* Assume just because you know something that everyone else does.	* Disseminate information more than once, especially if it is important. Unfortunately, repetition sometimes equals importance.

FIGURE 2. YP mindset.

versal technique" is a common exercise conducted in quality training sessions. The premise is to foster new ideas on how to improve a process. For example, the initial question might be, "How can we reduce manufacturing defects?" From there, the reversed question would be "How can we increase manufacturing defects?" The group would brainstorm ways to do this. After about 10 minutes, they would review their list of ideas for increasing manufacturing defects, and they would apply those ideas to the initial question of decreasing defects. The key is to look for bits and pieces that make sense. Some of the ideas don't make any sense and are dismissed. Many, however, will be quality ideas.

Never assume that someone will resist new technologies or methods. By doing this, you create an excuse immediately for why something cannot be done. It is more important to gauge what needs to be done to develop these new skills and then to take the proper steps to set that plan in motion.

ACTION ITEMS

Some of the most beneficial tools available to a manager are the use of action items and action plans. Such plans help track specific tasks and resources while providing information about effort required, due dates, completion dates, and related costs. Formalizing and documenting these plans as much as possible is appropriate for a YP manager.

Creating a knowledge base of information should be a priority. It is difficult to move forward and plan for the future without fully understanding where everyone is right now. The task of accumulating knowledge that resides in people's minds is a difficult one that requires much effort on everyone's part. However, the end result is worth the effort. It helps considerably with the problem of acquired knowledge being lost when a person leaves the group. More times than not, the way something is done is not documented and happens because this person is conscientious and just does it well. Alleviating this and having the information to pass on to the next person will decrease the time spent in the learning curve. The key is to document as much as possible. Anyone who works with the ISO-9000 certification process can appreciate the necessity of documentation.

Another important facet is learning from the process of gathering information. By making this as interactive a process as possible, a YP can gather a great deal of information from older subordinates. This structured approach facilitates the free exchange of information.

In a similar fashion, it is also important to understand the background and histories of the subordinates. Take the opportunity to create histories, including work, projects, and assignments. The interactive communication leads to a better awareness of the person's interests, capabilities, and strengths, as well as areas of growth potential. With this information, delegation becomes more effective and resource planning becomes more productive.

Documenting action plans is imperative for successful managers. With the pace that business sets today, things are over before you know it. Technology advancements, the prevalent run-and-gun attitude, and the get-up-and-run approach to implementation all dictate maintaining a lead on all the business dogs chasing you. An action plan can be something as simple as a handwritten document or a spreadsheet, or can even be a specifically designed computer program. Activities can be easily tracked, evaluated, and if need be, adjusted. These plans should direct and lead professional growth and can be used to encourage participation in professional and developmental societies.

CONCLUSION

To be successful, managers must possess some common traits. The abilities needed by YPs to supervise older subordinates form a subset of these common management traits. All are useful for establishing the mindset of YPs as supervisors and managers and may help YPs establish a management philosophy.

REFERENCES

1. Covey, Stephen R., *Principle Centered Leadership*, Simon & Schuster, New York, NY, 1990.
2. Gallagher, Rich, *Viewpoint: Three Secrets of Excellent Customer Service*, Service News, United Publications, May 1998.

3. Joiner, Brian L., *Fourth Generation Management*, McGraw-Hill, Inc., Multiple Cities, 1994.

ABOUT THE AUTHOR

Sean Lockhead is product support group manager at Kaman Industrial Technologies in Tonawanda, New York. For more information, write to Mr. Lockhead at Kaman Industrial Technologies, 245 Cooper Avenue, Tonawanda, NY 14150.

Trends in Employee Benefits

G. F. McCLURE

ABSTRACT

The average cost of employee benefits stood at 41.3 percent of payroll in 1996. Employers, recognizing the growing number of dual-worker households, are providing flexible benefit packages more often. This paper discusses key decisions facing workers as they shape individual packages, as well as current legislation affecting benefits.

INTRODUCTION

While the employer includes the cost of vacation and time off, Social Security, Medicare, workers' compensation and unemployment insurance costs as employer-paid benefits, the employee thinks more in terms of life and health insurance, pension plan, 401(k) plan, and a variety of fringe benefits. Being knowledgeable about what the employer offers helps employees select intelligently the benefits that would be most valuable to them.

THE MENU

A checklist of potential benefits could include:
- Educational reimbursement
- Preferred rates for personal travel
- Parking and mass transit subsidies
- Van pool
- Moving expenses
- Relocation services
- Matching gifts to charities
- Anniversary programs

George F. McClure, Member, IEEE-USA Engineering Employment Benefits Committee

- Severance pay
- Outplacement center/assistance
- Product discounts
- Legal services
- Profit sharing
- Overtime pay
- Physical exams
- Fitness center
- Employee assistance
- Sickness/accident insurance
- Short- and long-term disability insurance
- Dental care
- Vision care
- Sick days
- Health screenings/seminars
- PPOs and HMOs
- Discount entertainment tickets
- Child care
- Dependent care resource/referral
- Flextime
- Sabbaticals
- Spouse job search
- Parental leave
- Home purchase assistance
- Accident insurance for spouse, children
- Psychiatric or marital counseling
- Credit union
- 401(k) or 403(b) salary reduction savings plans
- Matching savings plans
- Thrift compensation bonus plans
- Deferred compensation bonus plan
- Life insurance
- Supplemental insurance
- Vacation; paid holidays
- Financial counseling; tax return preparation
- Survivor income benefits
- Dependent life insurance
- Funeral leave
- Jury duty/military leave
- Reimbursement accounts
- Flexible spending accounts
- Pension plan
- Salary deferral plan

- Simplified Employee Pensions
- Supplemental retirement benefits
- Post-retirement counseling

WHERE BENEFIT DOLLARS ARE SPENT

Spending for employee benefits in 1996 was divided in the following proportions, according to the U.S. Chamber of Commerce's *Employee Benefits 1997 Edition*:

- 25 percent—Vacation and other time off
- 23 percent—Health insurance and other health-related benefits
- 21 percent—Social Security, Medicare, Workers' Comp., Unemployment, and other required benefits
- 15 percent—Employer contributions to retirement and other savings plans
- 9 percent—Paid rest periods
- 6 percent—Education assistance, product discounts, other miscellaneous benefits
- 1 percent—Employer contributions to life insurance

The benefit cost for large firms went down while the cost for small firms (under 100 employees) increased. This change is attributable to the decline in downsizing and restructuring, which reduced costs for big companies, and to growing competition for labor among small companies, which has pushed up their benefit costs as they have worked to attract and keep employees. Severance pay, payments to defined benefit pension funds, and retiree health insurance costs all went down in 1996, compared to 1995 levels. The trend toward managed care and transferring medical costs to employees has reduced health care costs to just under 10 percent of payroll, making vacations, holiday pay, and other time off the biggest benefit item. Workers' health insurance payments grew to 1.7 percent of payroll in 1996.

The 802 firms responding to the survey employed 2.3 million full-time equivalent workers, making this survey the largest of its kind.

More companies are contributing to employee retirement plans, but they are contributing less than before, according to the survey. About 87 percent of companies surveyed contributed to a savings plan, pension fund, profit-sharing plan, or similar program, but the amount paid into those plans dropped to 6.3 percent of payroll in 1996, down from 7.5 percent the year before.

Benefit costs are highest in the northeast (44.8 percent of payroll); benefits in the eastern north central region and in the west were 40 to 41 percent of payroll respectively, while in the southeast the figure was 39.9 percent.

CURRENT BENEFITS ISSUES

Employee benefit issues figure largely in legislation in the 105th Congress, as bills introduced place additional mandates on employers. The Republican-sponsored "Patient Access to Responsible Care Act" (PARCA, H.R. 1415/S. 644) provides for redress of employee problems with employer-provided health care. Many insurer and employer groups, including the U.S. Chamber of Commerce, claim the bill's 300 new requirements would force premiums up by 23 percent, on average, and therefore reduce the level of health coverage that employers could buy for their employees. The Health Benefits Coalition for Affordable Choice and Quality claims to represent three million employers providing health care coverage for more than 100 million workers and their families. It is opposed to PARCA and to the Democratic leadership's "Patients' Bill of Rights Act" (H.R. 3605/S. 1890). Both bills would give broader rights to patients to sue health plans for failure to provide necessary care than is afforded under the 1974 Employee Retirement Income Security Act of 1974 (ERISA). In 1997, the Advisory Committee on Consumer Protection and Quality in the Health Care Industry called for the establishment of two panels to set and monitor standards of health care quality, one on government and one in the private sector. This presidential advisory committee could not agree on enforcement, however, and stopped short of advocating patients' rights to sue. The president, however, ordered federal compliance with the "bill of rights," covering all beneficiaries of federal health insurance programs, including Medicare, Medicaid, and health plans for federal employees and retirees.

In rebuttal to the patients' rights initiative, the industry coalition claims that every one-percent increase in employer health costs forces 200,000 to 400,000 Americans to lose their health insurance. If legislation exposing employers to lawsuits over health care were to pass, the industry group would advise its members to stop providing health care coverage to their employees. Small businesses and their workers would be hardest hit. (See the References list for further information resources.)

Another bill would give employees the right to choose whether they want compensatory (comp) time off or pay for overtime hours worked. The bill, amending the Fair Labor Standards Act, would permit a worker to choose 1.5 hours of comp time for every hour of overtime worked. Workers could accrue up to 160 hours of comp time in the House bill and up to 240 hours in the Senate version. The choice must be voluntary for the employee; any coercion by the employer would be penalized. Labor unions oppose the move, claiming some employers would use the option as a way to deny overtime pay.

FLEX PLANS

Flex plans or cafeteria plans are becoming more prevalent, as dual-income married couples try to coordinate their benefits packages to avoid duplication. Examples of choices might be:

- Any of several medical plans, with different combinations of deductibles and copayments
- Dental coverage

- Vision care
- Dependent life insurance
- Choice of disability plans
- Life insurance coverage for employee
- Accidental death policy
- Pretax spending accounts.

The pretax spending accounts can be used to pay deductibles and copayments. The employee sets the level in anticipation of the amount of health care that is likely to be required during the year. Any money in the pretax spending account not used for health care expenses during the year is lost.

Medical Savings Accounts (MSAs) are a variation on the pretax spending account. Used in conjunction with a high-deductible health insurance policy, the MSA is used to pay the deductible when health care is actually required. Any money in the MSA not spent during the year rolls forward to the next year and ultimately to an IRA at retirement. This plan would be useful to young, healthy workers. IEEE's Financial Advantage program now offers an MSA in connection with high-deductible health insurance plans.

In the Fiscal 1999 budget, the administration has proposed that laid off workers as young as age 55 be given the opportunity to enroll in the Medicare system at a premium that would fully cover their cost. This might find an analogy in the COBRA entitlement, whereby employees can continue their employer-provided health care for up to 36 months after termination by paying the full premium cost themselves. Critics argue that Medicare's Hospital Insurance component is already in imminent danger of default, as expenditures exceed income and deplete reserves. Medicare spending cuts made last year extended solvency of the Medicare trust fund from 2001 to 2007. In order to preserve the program, reductions in care covered by Medicare may mean that the laid-off workers would be better advised to seek a managed care plan that would accept them, with the layoff being the triggering event that avoids the "adverse selection" that worries insurers—that is, that people voluntarily join a health plan only when they know that something will need medical attention shortly.

The best program to have for health care is the one that the Congress has for itself—the Federal Employee Health Benefits Program, with nearly 10 million enrollees nationwide. Consumer satisfaction with that program runs 87 percent for those in fee-for-service plans and 85 percent for those in HMOs. A total of 380 health plans nationwide participate in FEHBP. Typical premium cost is around $2500 per year for singles, $5500 per year for family coverage. Retired military, who are finding fewer military medical facilities available to them as bases continue to close, are lobbying for admission to FEHBP.

The Self-Employed Health Fairness Act of 1997 (H.R. 876) has been introduced to make the medical insurance premiums of self-employed workers fully tax deductible for tax years after 1996. Under present law, the gradual increase in the portion deductible does not reach 100 percent until 2007.

PENSION ISSUE

The assets in defined-benefit pension funds, to which employees do not contribute, are the property of the employer administering the fund until the funds must be paid out to fulfill pension claims. Hughes Aircraft had a contributory defined-benefit fund until 1991, when it switched to a two-tier plan. New participants made no contribution and were entitled to fewer benefits on retirement than the long-term employees. Older, non-retired employees could choose between contributory and non-contributory options. In 1992, five retired employees initiated a lawsuit over a $1.2 billion surplus in the contributory pension plan that they charge is being used for the benefit of new employees. At this writing, the U.S. Supreme Court has agreed to hear the case, which possibly could affect the 33 million American workers and retirees who participate in defined-benefit plans.

BIBLIOGRAPHY

1. *Employee Benefits Plain and Simple,* by James M. Jenks & Brian L.P. Zevnik, Collier Books, 1993
2. "Company-Paid Costs Decline Slightly," by Stephen Blakely, Nation's Business, Feb. 1998
3. Weekly Standard, Feb. 2, 1998, p. 5 [Health Benefits Coalition advertisement]
4. National Organization of Physicians Who Care, http://www.pwc.org
5. The Health Benefits Coalition for Affordability Choice & Quality, press release dated April 27, 1998. http://www.hbcweb.com/prhbc/prhbc16.htm
6. Patient Choice and Access to Quality Health Care Act of 1998 (H.R. 3547), http://www.pwc.org/hr3547.htm
7. "Workers' Desire for Time Off," by David Warner, Nation's Business, December 1997
8. "Medicare, Welfare Raises Proposed," Facts on File, February 5, 1998
9. "Medicare, Medicaid Spending Slowed," Facts on File, May 8, 1997
10. "A Health Care Plan Most of Us Could Buy," by Eric B. Schnurer, The Washington Monthly, April 1998, pp. 20–25
11. Self-Employed Health Fairness Act of 1997 (H.R. 876), http://thomas.loc.gov
12. "How HMOs Decide Your Fate," U.S.News & World Report, March 9, 1998, pp. 40–50
13. "High Court to Hear Hughes Aircraft Dispute Over Pension Claims," Dow Jones Newswires, April 27, 1998
14. Employee Benefits, 1997 Edition is available for $35 in printed form; call 1-800-638-6582 weekdays between 8:30 a.m. and 6 p.m. eastern time. In Maryland call 1-800-352-1450.

ABOUT THE AUTHOR

George F. McClure, an IEEE Life Fellow, is a member of the IEEE-USA Engineering Employment Benefits Committee, the IEEE Individual Benefits and Services Committee, and the IEEE Insurance Committee. In addition, he is serving as PACE Coordinator for Region 3. He retired from Martin Marietta Aerospace in 1993, where he worked in communications systems engineering and research and technology for more than 30 years. Before joining Martin Marietta Aerospace he worked with Radiation, Incorporated, and the U.S. Navy.

Mr. McClure received a bachelor's degree in electrical engineering and a master's in engineering from the University of Florida. He can be reached by e-mail at g.mcclure @ieee.org.

Is Tax Relief Real?

G. F. McCLURE

ABSTRACT

The Taxpayer Relief Act of 1997 added complexity to the income tax code in the name of simplification. An assortment of targeted tax cuts was included among 225 new tax provisions. In this paper, implications of the tax changes are explored, the opportunity to improve retirement security presented by the new Roth IRA are reviewed, prospects for further tax changes are discussed, and the pros and cons for future tax reform are weighed.

INTRODUCTION

A variety of targeted tax cuts and incentives for parents, students, investors, and savers were included among the provisions of the Taxpayer Relief Act of 1997, enacted as Public Law 105-34 on August 5, 1997. New complexity and an opportunity for reduction were introduced into the computation of the capital gains tax. A new back-ended IRA promises future tax-free retirement income. The required Alternative Minimum Tax computation is spreading from upper-income taxpayers using tax shelters to mid-level dual-income taxpayers. More filers are subject to the marriage tax penalty than ever before, as more spouses move from homemaker and child-care roles to jobs in the workplace. Critics charge that the tax breaks were designed so that most of the cost comes after 2005. While the five-year cost is $95 billion, the ten-year cost is $275 billion.

THE BREAKS

The tax bill was developed with the goal of balancing the federal budget in five years.

George F. McClure, Member, IEEE-USA Engineering Employment Benefits Committee

A net cut in tax revenue of $95 billion over five years resulted from offsetting $151 billion in tax breaks with about $56 billion in new taxes on foreign air travel and tobacco. Principal beneficiaries are:

EDUCATION

- A $400 per dependent child tax credit, equivalent to a $1425 deduction for a taxpayer in the 28-percent bracket. This credit, for children up to age 16, rises to $500 from 1999. It is, however, phased out for joint returns over $110,000 per year, $75,000 for individuals, or $55,000 for married filing separately. The credit is reduced by $50 per $1000 of income above these limits. The existing deduction for a dependent, now $2650, is independent of this credit.
- An education tax credit of $1500 for each of the first two years of college enrollment, termed the Hope scholarship. The student must be enrolled at least half time and must not have a felony drug offense conviction during a year in which the credit is claimed.
- A lifetime learning credit for any taxpayer for any tax year, up to $1000 (20 percent of qualifying education expenses up to $5000) through 2002, or $2000 (20 percent of $10,000) thereafter. Both education tax credits go away when adjusted gross income exceeds $50,000 for a single return or $100,000 for a joint return, and start to reduce for incomes 20 percent lower. No credit is allowed for marrieds filing separately.
- Tax deductibility of interest paid by taxpayer on student loans, up to $1000 for 1998, increasing in $500 increments to a maximum deduction of $2500 in 2001 and thereafter. The deduction is available only for the first 60 months of loan repayment and phases out for incomes between $60,000 and $75,000 (joint returns) or $40,000 and $55,000 (individual returns). It is not available for marrieds filing separately or for dependents.
- Taxable distributions from IRAs to pay education expenses beginning in 1998 will not have the 10-percent penalty normally applied for early withdrawal, provided the education is provided by a qualified higher education institution and is for the benefit of the taxpayer, spouse, dependent, or grandchild.
- Education IRAs permit $500 per year per child to be put away for future qualified higher education expenses, provided the child has not reached age 18. Contributions are not tax-deductible but growth of the account is not taxed on withdrawal. The fine print limits the full advantage of this provision to joint filers with incomes below $150,000 and single filers below $95,000, and it is not available at all for incomes above $165,000 and $105,000 respectively.
- Exclusion from taxable income of employer reimbursement for undergraduate education expenses, not directly job-related, up to $5250 per year, during the period January 1997 through May 2000. No comparable exclusion is available for graduate-level class expenses. Reimbursement for education that is directly job-related is deductible without limit.

RETIREMENT SAVINGS INCENTIVES

In 1997, for the first time, a non-working spouse could contribute up to $2000 to an IRA instead of the earlier limit of $250, if the couple file a joint return, if the wage income of the couple covers it, and the spouse is younger than age 70-1/2. The rules for tax deductibility of the IRA contributions of both spouses have been complex. If either is participating in an active retirement plan at work (such as a pension plan or a 401(k) plan), then there is a phase-out for deductibility, beginning at $40,000 modified adjusted gross income, or MAGI (modified in the sense that untaxed income such as from municipal bonds must be added into the total). At $50,000 MAGI, all deductibility disappears. For single taxpayers, the threshold is at $25,000.

For 1998, the phase-out threshold amounts are increased to $50,000 and $30,000, respectively. These caps will rise to $80,000 for joint returns in 2007 and thereafter, and to $50,000 for single taxpayers (not marrieds filing separately) in 2005 and thereafter. If only one spouse is covered by a retirement plan, then the other spouse will be allowed tax deductibility on IRA contributions, provided the MAGI is below $150,000.

WITHDRAWAL PENALTY REPEALED

The 15-percent penalty formerly levied on large retirement plan withdrawals—more than $160,000 from an IRA or $800,000 from a company plan—had been suspended for three years. Now it has been repealed.

ROTH IRA

A new form of IRA is available starting in 1998—the back-ended or non-deductible IRA, also called the Roth IRA in honor of its father, Sen. William Roth, Chairman of the Senate Finance Committee. While the contributions (still capped at $2000) are not tax-deductible, the growth of funds in the IRA is not taxed on payout, provided they have been in the account at least five years and the account holder is at least 59-1/2 years old. Exceptions allow withdrawals to buy a first home or to pay college bills without penalty. Contributions are limited if joint income exceeds $150,000 or singles income exceeds $95,000, and are totally banned at $160,000 and $110,000 respectively.

Other features of the Roth IRA are elimination of the 70-1/2 age limit for making contributions, of the requirement to take the money out on a schedule (because no tax is to be paid), and of the ability to include Roth IRA assets in an estate to be passed by inheritance.

A one-time feature, intended to generate some tax revenue, is the option to move IRA assets from a conventional IRA to a Roth IRA and to pay the tax due over a four-year period. This feature is available only to taxpayers with adjusted gross incomes below $100,000 in 1998, and cannot be used by married taxpayers filing separately. It should be exercised late in the year, when it is certain that the income limit won't be exceeded, because penalties exist for improper conversions, in addition to making the whole tax bill payable in one year.

IRA analyzer software is available from several sources. This software helps evaluate the desirability of switching from a conventional to a Roth IRA. Web sites for Charles Schwab (www.schwab.com), Salomon Smith Barney (www.smithbarney.com) and Fidelity (www.fidelity.com) have on-line analyzers; T. Rowe Price (www.troweprice.com) has a discussion on conversion at its web site and offers a disk-based analyzer for use on your PC.

CAPITAL GAINS TAX BREAK

Some sellers of long-term holdings having a profit will benefit from changes in the capital gains tax structure, but at the expense of complexity. For long-term capital gains, previously defined as gain on property held longer than a year and taxed at 28 percent, there are now capital gains tax rates ranging from 28 percent down to 8 percent—including 25, 20, 18, and 10 percent—depending on the type of asset, the holding period, and when it was sold. The lowest rate, 8 percent, applies to capital assets held longer than five years by those in the 15-percent tax bracket who make the sale after the year 2000. For most people, the 20 percent rate applies to securities held more than 18 months, the 28 percent rate to assets held more than 12 months, and the 18 percent rate to assets held more than five years (but sold after 2005). The gain on depreciable real property held more than 18 months is taxed at 25 percent.

SALE OF RESIDENCE

As before, homeowners selling their home at a loss cannot deduct the loss from their income for tax purposes. However, up to $500,000 of current and deferred profit on sale of a principal residence is now free from tax for joint filers starting after May 6, 1997 ($250,000 for single taxpayers). This tax break is available every two years, provided the taxpayers have lived in the home for the previous two years. So one sale could be the home in town and a later sale could be the vacation home that is used as a primary residence after retirement. Any gain above the tax-free limit is taxed at the new rates.

ESTATE TAXES

The old $600,000 exclusion from estate taxes moves to $625,000 in 1998 and gradually increases to $1 million by 2006. Had the $600,000 limit, set in place in 1987, been indexed for inflation all along, it would be now be $854,000. For family-owned businesses, $1.3 million can escape tax in transferring control of the business after a death in the family. A married couple owning the business can both use the special exclusion, increasing the tax-free amount to $2.6 million, with careful estate planning.

The $10,000 limitation on gifts excluded from gift taxes (per donor to each recipient) is indexed for inflation in years after 1998, in increments of $1000. Likewise, the generation-skipping transfer (GST) tax exemption of $1 million (for gifts to grandchildren or great-grandchildren) is indexed for inflation in years after 1998, in increments of $10,000.

HOME OFFICE DEDUCTION

Recognizing that many businesses that are not conducted exclusively in a home office nonetheless must have a location where records are kept and the business is managed, the new law, starting in 1999, liberalizes the application of the home office deduction. This change may be worth $2.3 billion to small-business owners who qualify, over a period of ten years. Consultants are among the workers affected. There are two criteria:

- The office is used by the taxpayer for administrative or management activities;
- The taxpayer does not perform substantial administrative or management activities anywhere else (some paperwork could be done elsewhere as long as most of it was done in the home office).

When the home office is established, the taxpayer can deduct transportation costs between the home office and clients' places of business. Good record-keeping is essential to prove the validity of the deduction to the Internal Revenue Service. Home office deductions will continue to be a frequent audit item.

HEALTH INSURANCE PREMIUMS

Beginning in 1997, self-employed individuals can deduct from taxable income more of the cost of their health insurance premiums. The deductible fraction in 1998 and 1999 is 45 percent, gradually rising to 100 percent in 2007 and thereafter. The part of premium cost that is excluded can be deducted under itemized medical expense, to the extent that the total exceeds 7.5 percent of adjusted gross income.

The cost of qualified long-term care insurance premiums and services can also be deducted as itemized medical expenses, whether or not the policyholder is self-employed, beginning in 1997.

ALTERNATIVE MINIMUM TAX

The Alternative Minimum Tax (AMT) was created in 1986 to ensure that some taxpayers benefiting from tax shelters and large deductions ("tax preference items") paid at least some income tax. For the AMT calculation, tax shelter deductions are omitted and items such as tax-free income from municipal bonds are added back in to calculate a second tax. The tax rate is 26 percent on income up to $175,000 after an exemption of $45,000 for joint filers ($33,000 for single filers) is deducted. The taxpayer pays the higher of the two taxes—the regular or the AMT. In 1997, 600,000 filers were subject to the AMT, but that number should jump to more than 6 million by 2006. The reason for this jump is that the thresholds are not indexed, and have remained unchanged since 1986. Congress considered AMT indexing in the Taxpayer Relief Act of 1997, but it did not appear in the final version. Most tax preparation software will do the calculations automatically and invisibly, but a preparer with calculator, pencil and paper should be armed with extra coffee to get through the added calculations. Taxpayers claiming the new child tax credit and the Hope tuition tax credit may find that they have to run the AMT calculation.

UNDERPAYMENT PENALTIES

For 1998 the exemption from penalty for underpayment of taxes (tax due minus withholding and estimated tax payments) increases to $1000 from $500. To avoid underpayment penalties taxpayers whose income is all from employment can adjust their withholding through changes in the W-4 to be sure that 90 percent of the current year's tax liability is withheld. Some taxpayers routinely request over withholding, so that they get a refund at tax time, but this means that they are providing an interest-free loan to Uncle Sam for up to a year. A better plan is to shoot for the 90 percent withholding level and have the use of the excess for growth during the year—possibly to fund the IRA earlier. Taxpayers with self-employment and/or investment income can file quarterly estimated taxes for the same effect. If the current year's income is uncertain, paying 100 percent of the previous year's tax liability will avoid penalty.

Withholding is assumed to be spread evenly through the year, so if you see a greater tax liability than you had been planning for accruing toward the end of the year, you can step up your withholding for the rest of the year to avoid penalty.

Make quarterly estimated tax payments promptly when due, to avoid growth of underpayment penalties. Rather than applying overpayment from previous taxes to next year's estimated tax, write a separate check. This avoids problems that may occur if an error in the previous return reduces the amount of refund and therefore reduces the amount you will be able to apply to the estimated tax.

Future Tax Cuts?

As more married couples become dual earners, the marriage effect on the tax due becomes less palatable. Depending on the relative incomes of the pair, the tax due may range from 18 percent more than would be due if the couple were single filers rather than married to somewhat less. Bills have been introduced to mitigate the effect of the marriage penalty; complete elimination could cost the Treasury between $18 billion and $30 billion per year in lost revenue. Expect the "family-friendly" Congress to take some action here; meanwhile some writers are extolling the financial advantage of divorce, IRS-style.

Another bill would exclude from tax the first $400 of interest and dividend income that couples receive from savings and investments ($200 for singles). This would restore a feature that was available years ago.

Tax Reform?

Don't count on it. A flat tax calculated on the back of a postcard sounds convenient, but to keep the rate low, deductions for homeowners, charitable contributions, health care, and education would have to be eliminated. Savers who rolled their regular IRAs into Roth IRAs and paid the tax would be chagrined to learn that they might have cashed out at a lower tax rate later.

The Congressional Budget Office (CBO) has studied a variety of tax reform proposals, including Dick Armey's flat tax (17 to 20 percent on wages, salaries and pensions only); Dick Gephardt's 10-percent tax (the only deduction is for mortgage interest; none for IRAs or 401(k) plans), in which three-quarters of taxpayers pay no more than 10 percent and the rest are in higher brackets up to 34 percent (cut from the present 39.6 percent); a national retail sales tax (everything taxed when purchased, even if payments spread over years through financing); a value-added tax (paid by the producers and distributors; invisible to consumers); and a consumption tax that allows funds to grow tax-deferred until spent or "consumed." The consumption-based tax was promising, and a future study will be devoted to it, but the CBO noted that it would have a redistribution effect on the tax burden. A choice in filing would be given the taxpayer under a proposal from Stephen Moore of the Cato Institute, who noted that Armey's flat tax (which Moore helped develop) covered wages only, not unearned income (rents; royalties; capital gains on stocks, bonds, and stock options; dividends and interest) and therefore heavily favored the wealthy. Cato proposes that the choice be between the present system or a 25-percent flat tax with no deductions but a credit for payroll taxes paid (Social Security and Medicare). Given the Congress' penchant for using the tax code for social engineering, and the difficulty of effecting a phase-in and transition from the present tax code to another one, these alternative taxes don't seem to have bright prospects.

BIBLIOGRAPHY

1. Taxpayer Relief Act of 1997, Public Law 105-34, August 5, 1997

2. Personal Tax and Financial Planning Guide 1997, American Association of Individual Investors

3. Tax Hotline, January 1998

4. "Are You a Winner?", US News & World Report, August 11, 1997

5. "J.K. Lasser's Your Income Tax, 1998"

6. Statistical Abstract of the United States, 1997, Table 752, Consumer Price Indexes

7. Economic Indicators, February 1998. Council of Economic Advisors [current CPI data]

8. The Schwab IRA Analyzer, http://www.schwab.com

9. Retirement Planning Aids by T. Rowe Price, http://www.troweprice.com/newira/analyzer.html

10. "Should You Convert IRA Assets Into the Roth IRA?" http://www.troweprice.com/newira/convert.html

11. Fidelity IRA Evaluator, On-line Retirement Planner, and Retirement Planning calculator, http://personal32.fidelity.com/retirement/buildassets/

12. "Easier Deductions for Home Offices," Joan Pryde, Nation's Business, March 1998

13. "An Arcane Levy Extends Its Reach," Joan Pryde, Nation's Business, November 1997

14. "The Economic Effects of Comprehensive Tax Reform: a CBO Study" Congressional Budget Office, July 1997

15. "Forsaking the Flat Tax," Bruce Bartlett; Washington Times, December 11, 1997

16. "Tax cuts? Funny you ask," James J. Glassman, US News & World Report, January 19, 1998

17. "Facts on File," July 31, 1997

ABOUT THE AUTHOR

George F. McClure, an IEEE Life Fellow, is a member of the IEEE-USA Engineering Employment Benefits Committee, the IEEE Individual Benefits and Services Committee, and the IEEE Insurance Committee. In addition, he is serving as PACE Coordinator for Region 3. He retired from Martin Marietta Aerospace in 1993, where he worked in communications systems engineering and research and technology for more than 30 years. Before joining Martin Marietta Aerospace he worked with Radiation, Incorporated, and the U.S. Navy.

Mr. McClure received a bachelor's degree in electrical engineering and a master's in engineering from the University of Florida. He can be reached by e-mail at g.mcclure@ ieee.org.

Networking: Multi-Level Marketing Yourself

D. J. PIERCE

You have all heard of multi-level marketing schemes. The power of these schemes is the large number of salespeople selling for you. In all multi-level marketing schemes, the people at the upper levels receive the greatest benefits. Wouldn't it be nice to apply the multi-level marketing principal to selling yourself and your skills?

You are your own best salesperson. Relatives, friends and people who are familiar with your capabilities make excellent salespeople for you, too. But your relatives and friends are limited in number. Your greatest marketing "team" expansion capabilities are found with people who are familiar with your capabilities. The opportunity to expand this group depends on your ability to network.

Everyone has networking experience. We all began networking as children; each time we met a new kid on the school playground we had an opportunity to expand our network; we just did not realize we were networking. We continued building our network during high school and college; fellow students and teachers who gained respect for our abilities became valuable members of our network.

We all do a certain amount of networking at our job sites. Our playground has become smaller and our network opportunities are limited to our coworkers, supervisors, and the clients with whom we interact. Some people do not have the opportunity to interact with anyone but their coworkers and supervisors, as their jobs insulate them from interaction with clients. This limits their ability to expand their network.

We all have the ability to expand our network. Participation at church, with community service groups, or with business organizations, provides us with opportunities to add members to our network. These groups allow us to network with a variety of individuals with different backgrounds and interests. Some will become valuable network resources, while others will never need our services.

People give many reasons for avoiding networking opportunities. With the downsizing and outsourcing activity so prevalent today, some argue that they do not have the time or are too busy at work. Some do not know what to say or are shy. Others are afraid of being rejected and do not know how to approach strangers or have difficulty remem-

David J. Pierce, Senior Engineer, Brooks and Jackson Consulting and Forensic Engineers, and IEEE Region 5 PACE Chairman

bering the names of people they have just met. All of these are valid reasons, but they are also easy to overcome.

Participating in technical organizations provides us with the greatest opportunity to expand our networks. These organizations allow us to network with individuals who have the same technical interests as we do. In fact, participating in your technical organization is similar to fishing in a well-stocked pond; no where else can you find so many individuals involved in the electro-technology field. Such organizations also offer a comfortable environment in which to practice networking.

The results of a recent survey of engineers indicate that those who participate in organizations such as IEEE find new jobs much faster during downsizing periods. This is due in part to these individuals' ability to use their networks.

All of the jobs I have held in my life came in part from members of my network. When I was young, I got a job at a service station because the owner knew my grandfather. My first job out of college came through a contact I established with a utility company engineering director when I was a student leader of my university's IEEE student branch. I met my current boss at an IEEE meeting many years before I went to work for him. Some of the consulting projects my firm undertakes are offered to us by contacts we made at IEEE and other engineering organization activities.

Technical competency is required to expand your network. The larger your network, the more important it is that you maintain the highest level of knowledge in your area of specialization. Your reputation is a great asset, and it requires a great deal of work to maintain. Participating in IEEE-sponsored technical conferences and seminars provides two opportunities: the chance to enhance your technical skills through educational programs, and the environment to network.

Networking is not rocket science. It is easy and fun to meet new and interesting people. The biggest impediment we face in building our networks is creating a lasting positive impression with individuals we meet. Most people will not remember you unless they meet you several times or unless they have a *reason* to remember you.

I find that my active participation in IEEE activities creates positive impressions with the people I meet. Organizing and implementing programs is a great way to create lasting positive impressions. Well-organized and well-run activities display your ability to be a leader, a mover and a shaker. My network has grown because of the exposure I have received by participating actively in IEEE.

The main goal of networking is to develop relationships for future referrals. You must learn what you have in common with other people. Common interests are a good starting point on which to build a relationship.

You must learn to become comfortable meeting new people. Many of us are uncomfortable around strangers. After all, we were taught to mistrust strangers as kids. Some of us are afraid of being rejected by the people we meet because we do not have any common interests. Rejection occurs, but as you perfect your networking skills, it will occur less frequently.

You have no way of knowing who the people you meet may know. There is only one way to find out: introduce yourself and get to know new people. You will never know whether or not a person will be an asset to your network unless you take that first step.

The key to networking is to find out what the other person does and knows and what their interests are. You already know yourself; networking is meeting and establishing relationships with new people. Always try to make a good impression on the people you meet. Try not to appear pushy. First meetings rarely result in opportunities. Networking takes work, demands discipline, and requires multiple meetings; the process takes time, but it's worth it.

Practice networking at every opportunity, until it becomes natural for you. Become active in organizations that will allow you to talk and exchange ideas. When possible, take an active leadership role in these organizations. Sit with different people at meetings or at dinner.

Participate actively in meetings that make time for networking opportunities before, during, and after the formal sessions. Show up early and stay late. Talk with as many people as possible. Do not be afraid to introduce yourself.

Remember to bring business cards, pens, and paper to networking opportunities. Exchanging business cards is great, if you use them. Make notes on the backs of the cards you collect. Remember to return calls, and deliver on any promises to keep the ball rolling. Re-establish contact periodically with these new contacts.

It takes time to develop relationships. Always reintroduce yourself to people you have met before. You can also help others network by introducing them to people you know. By doing this, you will be perceived as a mover and a shaker with an extensive network of resources—someone to get to know.

Always remember to thank people who have helped you, and let others know what those individuals did for you. If someone has helped you, let members of your network know what a good job the individual did for you. If you have an opportunity or know of an opportunity, talk about it. The best way to enlist a member to your network is to alert them to opportunities they might have missed otherwise.

Know yourself. Understand your capabilities and areas of expertise. Think like a businessperson. Determine several key benefits you can offer and that you want remembered. Summarize your key talents and benefits in a short "speech." Use key phrases that describe what you do and are easy to remember. Think about how individuals you meet can benefit by joining your network.

Work on your communication skills. If possible, attend communications workshops at conferences or through university or community college. Learn to listen attentively. Remember, you are trying to learn about the person you are meeting; the best way to do that is to listen to them. Try to determine family and hobby interests to find out what you might have in common. When talking, use humor, if it is appropriate.

Dress appropriately for networking opportunities. Make eye contact with people, smile at them, and introduce yourself with a handshake. Ask people to repeat their names if necessary, and then use their names frequently during your conversations.

During your conversations, do not be pushy. Do not overstate your capabilities or brag. Do not dominate the conversation. Do not tell off-color or demeaning jokes. Do not criticize others or their capabilities.

And practice, practice, practice. Market yourself and the members of your network by using the multi-level marketing principle. You will know that you have been successful when you meet someone new who already knows you. Your world will shrink as your network expands.

REFERENCE

Soltesz, Peter A., The Why's and How-to's of Networking

ABOUT THE AUTHOR

David Pierce is a senior engineer and certified fire and explosion investigator at Brooks and Jackson Consulting and Forensic Engineers in Baton Rouge, Louisiana. He has extensive training and experience in industrial and commercial electrical system design, construction and maintenance; electrical utility plant design, construction, planning, and operation; and interpretation and application of electrical codes and standards including IEEE, NEC, NESC, OSHA, and others.

Mr. Pierce has developed special expertise in fire origin and cause determination, accident reconstruction, and electrical and electronic product failure analysis. He has investigated numerous fires and has been court-accepted as an expert in electrical engineering and fire cause and origin determination.

David is a veteran of the U. S. Army (1972–1975), receiving specialized training in aircraft navigation, control, and communication equipment maintenance and repair. He worked as an industrial electrician and construction supervisor with Noble Drilling and Diamond M Drilling companies. He received his bachelor's degree in electrical engineering from the University of Southwestern Louisiana in 1982.

Mr. Pierce is chair of the IEEE Region 5 Professional Activities Committee for Engineers (PACE) and is a member of the IEEE-USA PACE Regional Activities Committee. He served Region 5 as the East Area Chairman (1992–1996). He is the vice chair of the IEEE-USA Member Professional Awareness Conference committee, and is a member of the IEEE-USA Precollege Education Committee.

Mr. Pierce served as the branch chair of the University of Southwestern Louisiana's IEEE student branch. He is active in the IEEE Lafayette Section, serving in all offices and as section chair (1989–90). He served as the Lafayette Section PACE chair from 1989–1996.

Mr. Pierce is a member of Eta Kappa Nu, National Association of Fire Investigators, International Association of Arson Investigators, National Fire Protection Association, and the International Association of Electrical Inspectors. He received the IEEE Region 5 Individual of the Year award (1990), the IEEE-USA PACE Leadership award (1992), and the IEEE-USA Regional Professional Leadership Award (1996).

NETWORKING EXERCISE

Find someone who meets the following criteria and have them initial next to the item. You have 10 minutes to do this.

Find a person who has a master's degree or higher.
Find a person who has lived in Korea.
Find a person who speaks a foreign language.
Find a person who has two or more children.
Find a person who has international consulting experience.
Find a person who is a pilot or has flown a single-engine airplane.
Find a person who owns a house.
Find a person who owns a British car.
Find a person who has been inside the White House.
Find a person who owns a German Shepherd.
Find a person who will receive an IEEE-USA award.

Benchmarking Engineering Skills Against a Rapidly Changing Future

L. VAVRA

ABSTRACT

Based on the premise that rapid technological advances render engineering skills obsolete in an ever shorter time span, this paper describes a workshop designed to offer options for engineers to stay viable and employable. The workshop helps participants document development goals to tailor career options into a SMART format.

The presentation borrows techniques from the Total Quality movement, including vocabulary such as benchmarking, strategic planning and continual improvement. Interactive exercises using large-scale (8' × 3') templates, diagram flipcharts, and individualized worksheets are tools used in the workshop, and they enable individuals to leave with a plan in hand for benchmarking their engineering skills.

The workshop was developed and presented initially to engineers in the aerospace industry. It was designed to appeal to the creative mind, adapting problem-solving analysis to an engineer's career development.

INTRODUCTION

Following the Socratic approach to education, it is important to incorporate strategic planning into your list of priorities, in order to keep your skill sets current. This way of learning, strategic thinking with action planning, has proved effective in previous seminars given. Issues covered in the workshop include:

- Social, technological, economic, environmental, and political changes
- Identifying and taking action on opportunities to strengthen your effectiveness
- Continual improvement using such transformation tools as environmental scanning, issues management, vulnerability-opportunity assessment, and scenario planning.

If you want to learn how to get more out of yourself, your career and your field, then you are ready to take steps to increase your own current value in the work force. You

Lynn Vavra, Principal/Founder, the Millennium Management Group

are also committed to improving your future market value—your ability to advance and earn more. This workshop helps you make specific determinations relative to your value. Topics covered include:

- Your role in the process;
- Who is your customer?
- How do you measure quality?
- Benchmarking;
- Career process quality management;
- Strategic planning;
- Empowerment and individual development plans (IDPs);
- Motivation for growth; and
- Prognosis for growth in the engineering field.

BETTING ON THE FUTURE

Technological advances render skills obsolete in shorter spans of time. Engineers who fail to keep up with the rapid change in some specialties risk technological obsolescence. This could make them more susceptible to layoffs or, at a minimum, more likely to be passed over for advancement. This program uses collective and individual strategic planning to predict and offset the risk of obsolete skills.

THE PROGRAM: A THREE-STEP PROCESS

During the program, participants are taken through three distinct steps:

- *Identifying intervening environmental factors, such as technological advances, government regulations, the global marketplace, etc.* Employment of electrical and electronics engineers is expected to increase "faster than the average" for all occupations. Increased demand by businesses and government for improved computers and communications equipment is expected to account for much of the projected employment growth. Engineers keeping current in technological advances are best poised to take advantage of growth opportunities (U.S. Department of Labor's *1998 Occupational Outlook Handbook*).
- *Bringing these environmental factors into your sphere.* What does it mean to my (potential) employers and my work group? Most importantly, how does it affect me? Job openings resulting from job growth and the need to replace electrical engineers who transfer to other occupations or leave the labor force should be sufficient to absorb the number of new graduates and other entrants, making for good employment opportunities through 2006. The need for electronics manufacturers to invest heavily in research and development to remain competitive will provide openings for graduates who have learned the latest technologies.

Consumer demand for electrical and electronics goods should create additional jobs. Job growth is expected to be fastest in non-manufacturing industries, however, because firms are getting electronics engineering expertise from consulting and service companies more often. Mid-career engineers are looking more frequently toward joining consulting firms, since opportunities for consultants are growing (U.S. Department of Labor's *1998 Occupational Outlook Handbook*).

• *Creating an individual development plan and career goal.* How can I prepare *now* to offset the rapid changes of the future? I'm no psychic; what are my options? Emphasis is placed on the importance of strategic career goals (Individual Development Plans).

SUMMARY

Participants have a chance to interact, brainstorm and debrief. But what do they do with all of the information they received while going through the three-step process? Some work in small groups, some by themselves, to create their own personalized plans. They walk away with knowledge of how to plan strategically for continual career development, and they actually put pen to paper in developing an Individual Development Plan. They learn about the skills they need to create future career development plans, which will keep their skills base sharp and keep them ready to compete for future jobs.

WORKSHOP REFERENCES

Camp, Robert C., 1995. Business Process Benchmarking. ASQC Quality Press, 1995, ISBN 0-87389-296-8.

Carlzon, Jan. Moments of Truth. 1989 Harper & Row, NY ISBN 0-06-091580-3.

Department of Labor-Bureau of Labor Statistics, Occupational Outlook Handbook, 1998.

Grove Consultants International, 1996. Process Tools & Strategic Planning: Context Map. San Francisco.

Goldratt, E. and Cox, J. 1992. The Goal. ASQC Quality Press. ISBN 0-88427-061-0.

OTH On-Line, ed. J.L. Morrison, 1998. Focusing Our Organizations on the Future: Turning Intelligence into Action. http://horizon.unc.edu/horizon/online/html/5/1

Senge, Peter M. 1990. The Fifth Discipline. Doubleday Currency, ISBN 0-385-26095-4.

Scholtes, Peter, 1996. The Team Handbook. Joiner Associates, ISBN 1-884731-11-2.

Shelby, C., 1998, High-Tech Cheap Labor Shortage, Investor's Business Daily, 30 Mar 98.

Walton, Mary. 1989. The Deming Management Method. Mercury, London, ISBN 1-85251-067-6.

ABOUT THE AUTHOR

Lynn Vavra is founder/principal of The Millennium Management Group, a consulting firm specializing in contract employee development counseling in Southern California. She has extensive training and experience in employee development, training, management and personnel selection.

Ms. Vavra has more than five years of experience working on a contract basis in organizational career development. Clients include organizations in the government, aerospace, and entertainment industries. She received a bachelor's degree in communications/marketing from Southern Illinois University-Carbondale (1982) and a master's degree in counseling with a specialization in business, industry and government from California State University-Northridge (1995).

Ms. Vavra is a board member of the American Society of Training and Development's Los Angeles Chapter. She is active in the Alternative Dispute Resolution Program in Los Angeles, serving as a mediator in the city council's Days of Dialogue racial relations program. She has published articles in the International Personnel Management Association Quarterly and has written job search webpage articles for American Online's Digital City local site. She recently wrapped up weekly chatroom host duties on AOL, with topics centered on career issues. Ms. Vavra is a member of the California Career Development Association and the National Career Development Association. She can be reached at (213) 656-3688.

Career Management as Personal Marketing and Business Development

C. VOEGTLI

ABSTRACT

What's the best overall career advice I can give you? *You need to think like a marketing person.* Whether we are working as design engineers, group leaders, functional managers, project managers, or consultants, we are the product or service someone else is employing. Given the rise of outsourcing work to consultants and contractors and the increasingly acceptable mobility of engineers from company to company, we have to understand what our customers and "users" are looking for, make sure we can deliver, and make sure we even get a chance at the opportunity. Are you the most desirable person to hire? What attitudes and skills will make you the most "sellable" and ultimately successful throughout your career? How do you make sure you have access to all the available opportunities? This paper provides:

- A comprehensive picture of the skills you need to develop as you move through your career, with the very important context of why you need them (from the customer's viewpoint), and an integrated understanding of how presentation skills, technical expertise, meeting management skills, networking, business understanding, etc., can provide you with incredible career leverage;
- A method for coming up with a personal strategy for developing those skills; and
- Ways to continually "market" yourself—communicating and using your capabilities to maximize your opportunities, your success, and your overall career satisfaction.

INTRODUCTION: UNDERSTANDING YOUR CUSTOMER

On the engineering projects in which we participate, the primary function of a marketing or business development person is to understand the potential market and customers for the product. Who are these customers and where are they? What do these customers think they want? What problems do they have and how can the product solve

Cinda Voegtli, President, IEEE Engineering Management Society, and President, Emprend LLC those problems in a cost-effective and timely manner? Who are we competing against for this business or market share? What other factors will affect customers' purchasing decisions? What will make them become repeat customers? What will make them recommend the product to their friends? How do we form long-term relationships with customers and create new business opportunities?

Your customers are the people you work for and with on your projects. You have to understand what your customers and "users" are looking for, make sure you can deliver, and make sure they know you can deliver. What skills, outlooks, and mindset do you need to make sure you can compete successfully for the corporate positions or consulting assignments you want, and have access to all the exciting future career opportunities you could wish for? To market yourself and to develop new business opportunities for yourself continually, ask yourself these marketing questions:

Who are our customers and where are they? Your primary customers are all those for whom you work directly, such as your functional manager and the managers of any projects you are working on. They are potential customers when they are interviewing outside candidates for a job or are looking for people inside the company to staff their projects. They are current customers as soon as you're doing work for them (whether as an employee or a consultant).

Your customers also include those *with whom* you work on projects—anyone who uses the output of your work, for instance. These could be called "secondary" customers. They aren't doing the active hiring or selecting, but they can be very influential to your success. For instance, if you're a hardware engineer, one customer is the purchasing representative who takes your bill of materials and buys the components necessary to build prototype and production units. (Yes, you're a customer of his, too—he is providing a service by getting parts for your hardware. But he is a customer or user of the information you produce and you must understand his needs.) Another is the designer who will turn your schematics into a printed circuit board. If you're an applications engineer, one customer is obviously the engineer who uses your company's components or systems in his or her designs. Other not so obvious customers are the designers in your company who rely on the feedback you get from those users, and the marketing and publications people who need your help with creating applications notes that will assist and attract customer engineers.

The identity of your customers is important—you need to understand whom you're trying to serve and how to serve them well. The "location" of your customers is important for developing future opportunities for yourself.

What do these customers think they want? Functional managers and project managers might be looking for a particular mix of skills and experience or a specific technology expertise to fill a job or a project team member slot. Beyond that, they'll be looking for a set of "intangibles"—Do you have a cooperative attitude? Do you meet schedules? Do you understand how to work on a team?

Your other customers, such as the purchasing representative, hope that you will understand their job on the project and provide the input they need: give them an early list of long-lead time parts, produce an accurate bill of materials, use parts that have more than one source, etc. They too will care about the same intangibles as the other managers, including whether you have an attitude of respect for their work.

What problems or objectives does the customer have and how can the product solve or fulfill them in a cost-effective and timely manner? Now back up a moment from what the customer says they want in terms of individual qualifications. What are the overall objectives and related problems they are trying to solve? Project managers always have a goal: to get the project out on time, on budget, with high quality. Their particular problem may include overcoming technical risks on the project, completing it on time with limited resources, finding skilled enough team members, dealing with corporate politics, or combating low morale. The functional manager needs a competent engineer to do technical work; he may also need someone who would be good material for a project leader later, or even someone qualified to take his place as functional manager one day. He may need help mentoring young engineers, suggestions for department procedures that will reduce the number of coding bugs, or assistance in solving communication issues with other functional groups. The purchasing rep has to order parts early enough to support the overall project schedule; try to get good prices for the company; and avoid vendors who've caused the company problems in the past.

Note also that a marketing person has to consider more than just the problem the customer says he is trying to solve. What is the *underlying* problem? For instance, I was once asked to evaluate a medical project that was running late. The director expected that I would suggest a little project management "stuff" to get it back on schedule, but he did not want to hire a consultant to manage the project. He thought he just had a minor scheduling problem. But what I found was that the project requirements were still changing; there were several significant technical risks; the outside developers didn't plan to do code reviews, unit testing or system testing; and the software developer was very inexperienced. That director's problem was different and much bigger than the scheduling issue he wanted quick help on. I got an eight-month project management contract when I showed him what the real problem was and how I would help solve it.

Who are we competing against for this business or market share, and where might we have an advantage? Inside and outside a company you may be one of several people who are considered for a position or assignment. You may be up against a number of other job hunters for a particularly enticing start-up or management position. If you're consulting, you might end up competing formally against other consultants or firms. In all cases you might also be competing against fear and inertia—if it seems to be too risky or too much trouble to hire someone from outside, will the manager just make do with someone from inside?

Note that competition isn't necessarily the main point here. In today's climate in many areas there are plenty of interesting jobs to go around. Whatever the situation, your goal is to make sure that you always have excellent opportunities to choose from—the opportunities that matter most to you personally. Thinking like a marketing or business development person can help make sure that none of these possibilities are inadvertently closed to you. And for that matter, thinking this way will ensure that you're prepared if the job scene does get tighter or the competitive landscape more intense. One engineer I know has progressively built a portfolio of skills related to integrated circuit design—what he calls the irresistible skill set. He anticipated that as chips got more complex designers who understood logic design, circuit design, layout, and software would be more valuable than a single-skilled person. Now his resume stands out in the crowd and brings interesting opportunities his way.

What other factors will affect the customer's purchasing or recommendation decision? Whatever primary requirements the customer has, such as particular technical expertise, the end decision to employ you will include other considerations—some objective and some very emotional. The customer will look for objective proof that you can do the work—not just that your resume says you have a particular expertise, but also that you really understood the work and performed successfully. A hiring manager may discuss the work in depth with you to gauge your understanding, give you a technical interview with problems to solve, and call references. A project manager looking for a suitable team member may talk to people with whom you've worked in the past, both inside and outside the company.

Beyond the objective criteria, we all know from watching advertising that marketers rely on customers' emotions to pitch products—does the product imply that it will make the user beautiful, loved, successful, etc.? Similar emotional considerations will enter into your world as well: Does your potential customer feel good about working with you? Do they trust you to do a good job? Do you pass the "sleep test," meaning the hiring manager feels they will be able to sleep at night, even in the most intense project situations, with you on the job? Do they believe that you respect them? Does anything about you—personality, outward attitude, appearance, etc.—make them suspicious about whether working with you will be an enjoyable and successful experience?

What will make them become repeat customers? If you get what you expected from a product—your problem was solved, your goal was achieved without disappointment—chances are you'd buy that product again. Likewise, your customers are more likely to become repeat users of your expertise if you delivered good performance and solved their problem with a minimum of disappointment or headaches. Did you complete your work on time? Did you do a high-quality design and fix any "bugs" quickly? Were you easy to work with? Were you committed to supplying the information needed by other members of your project team? Were you responsive to requests for information and assistance? Do your customers trust you?

What will make them recommend the product to their friends? Positive answers to the "repeat customer" questions posed in the previous paragraph are the foundation for a product recommendation, but may not always result in a *proactive* recommendation of you to other potential customers. Think about your own experience with products: what does it take for you to run around bursting to recommend a product to a friend? For me the extra ingredients are: a product that absolutely delighted me with its performance, one that fully satisfied or exceeded my expectations; and a feeling of trust—that the product's creator cared about what I needed, did a good job creating a product for me, and charged fairly—and can be expected to do so again. If you can determine how to build this trust and beat those customer expectations, your future "business" will come to you. If you're a consultant, your next contract will be waiting in the wings. If you're a corporate employee, the manager of the next exciting project will be waiting to snap you up.

What new markets can you identify, create, and prepare for? Looking down the road, what work might you want to do later? What's a natural career path for your skills and interests? You can plan ahead, develop the necessary skills, and start identifying and wooing your future customers. What might be some unexpected but interesting possible branches to that path? You can think strategically and build a portfolio of skills and network of contacts that may later allow you to go in those interesting directions. Remember the IC design engineer mentioned earlier.

To illustrate how the above questions can come together to assist your career, let me put my customer hat on for a moment and give you an example. When I advise consultants or contractors on how to get hired by corporate managers, I give the following kinds of recommendations:

- When you find out about the job opening, find out who the hiring manager is and how to get hold of him directly. *(Know who and where your real customer is.)*
- Call and emphasize your interest in the specific position. *(You know that he may have received 100 other resumes for this position and you want to stand out.)*
- Inquire about the project schedule and let the hiring manager know that you are willing to be dedicated to their project full time. *(Understand that his main goal is to get the project done on time and show that you want to help him meet those dates. You know he's probably had trouble with consultants who have taken on multiple simultaneous clients and have missed schedules because of conflicting priorities.)*
- Have letters of commendation from past clients. *(Get that objective proof that will show the customer that you really can perform.)*
- Address the manager's concerns about using consultants. Commit to train the client on whatever you design, and commit to provide thorough documentation. *(Address his fear that consultants will hide and withhold knowledge in an effort to guarantee more business, or will do a sloppy documentation job out of lack of commitment or haste to move on to another contract.)*
- Be able to recommend others. *(Understand that this project manager probably has other open slots, a tight schedule deadline, and no time to read resumes and do interviews. He'd love to have other high-quality candidates recommended to him for an easy hire.)*
- Once hired, make sure you know how you'll get feedback from the manager on his view of your performance. *(You not only want to do a good job, you want to make sure the manager believes you're doing a good job so he'll use you again and recommend you to others.)*

- Eventually find out what other work this person is managing and what his company is doing. *(Form a relationship and turn this client into a long-term business partner.)*

THE SKILLS WE NEED AND WHY

Now that we've discussed the marketing/business development mindset you need to have to maximize your opportunities, let's delve deeper into the skills that will help you truly satisfy (and even delight) your customers.

First, think of the daily life of an engineer on a product development project:

- *Week 1:* You review Marketing's requirements specification and start thinking about how you could design this product.
- *Week 3:* You identify several possible design alternatives, some of which are more interesting technically than others.
- *Week 4:* You hold a high-level design review meeting to show your peers and a few managers the alternatives you've come up with.
- *Week 5:* The requirements and tradeoff process is taking a while and the team is getting impatient to be doing "real work"—technical detailed design.
- *Week 6:* You finish your detailed schedules for the rest of the project.
- *Week 8:* You start reporting status in a weekly report that the director may see.
- *Week 10:* You make a presentation to the field support group to show them how the product will work so they can start creating their installation and maintenance procedures.

Now, consider these same tasks, the related challenges you may face, and your choices in dealing with them.

Weeks 1–2: In studying the marketing specification, and working on design alternatives, you realize Marketing wants much more functionality than you think is possible by the delivery date they've specified. You can complain about yet another impossible marketing specification, or you can work a bit harder to understand the underlying business case, feature priorities, etc., and suggest tradeoffs and identify alternatives that could be developed in time. *You have a business-oriented perspective.*

Week 3: You realize that one of the design alternatives will seriously compromise both software performance and manufacturability. You can concentrate only on your technical concerns and let someone else worry about it, or you can raise the issues and recommend that this design alternative be abandoned (even though it is very interesting to you technically). *Your technical expertise is wider than just your primary area of responsibility.*

Week 4: You hold a design review meeting. You can hold an impromptu unfocused meeting that degenerates into design by committee, rambles around in endless conversation, delves too far into design details, wastes people's time, and/or never results in a decision, or you can plan ahead, issue advance technical material and an agenda, manage the meeting to be efficient and effective, cover the right level of detail, and make good decisions. *You have astute meeting management skills.*

Week 5: The requirements and tradeoff process have taken a while and the team is ready to get on with design. You can be a complainer, wishing management would just make up its mind, or you can realize that the tradeoffs are complex and that this decision process is necessary, and you can be a positive voice helping the other team members see its importance too. *You demonstrate a mature, constructive, flexible attitude.*

Week 6: You finish your detailed schedules for the rest of the project. You can take a quick cut at your schedules, complain about micro-management, and go on, or you can think back to previous projects, remember issues that arose due to poor planning, and adjust your estimates to be more accurate. You can go the extra mile to check the project dependencies the project manager has laid out and identify areas where a crossfunctional group isn't getting information from Engineering in time. You can make sure the project manager includes backup plans for the risky technical areas you're looking into. *You have a good understanding of project management issues.*

Week 8: You report weekly status that the Director sees. You can give a detailed data dump and let him find what he needs, or you can stop and think about what that director really wants and needs to know. Then you can work a little harder to produce a concise summary of progress, issues, and plans that will tell the director quickly what he needs to know, and concentrate his attention on where he can add value—helping solve your problems. *You have effective communication skills.*

Week 10: You make a presentation to the field force. You can treat the occasion casually and deliver an off-the-cuff presentation on your design, or you can prepare a nice presentation that captures their attention; gives a logical, interesting, well-organized overview of your design from their "user" perspective; and provides them with a summary they can also give to people in the field. *You demonstrate effective presentation skills.*

From these examples you can cull the skills you can develop and you can see why they're important from the customer's viewpoint:

Business understanding: You understand management's perspective of the issues. You can thus be a partner in solving their problems, rather than a potential source of problems or a complainer. Engineers are sometimes known for insisting on technical features or designs that aren't needed, will cost too much, or take too long to develop, simply because they're fun to design. Instead, you will be trusted to not just take a one-sided technical view. You can help these managers make the right decisions for the company. You are viewed as a valuable partner and resource.

Broad technical expertise: You can contribute to the success of an effort outside your core area of expertise and responsibility. Other functional groups (your secondary customers) trust you to look out for their interests as you design. The more people there are like you on the project, the more confident the managers are that risks will be understood, issues managed, and success achieved.

Mature, constructive, flexible attitude: The ability to deal with ambiguity, especially in the front end of a project, is one of the biggest differentiators I notice as a manager. If you also proactively help your team members do the same, you will be viewed as a leader as well, and will be competed for by project managers everywhere!

Meeting management: You understand that good meetings are simply a result of discipline in planning and execution and attention to normal "people" issues. You realize that the investment to get good at these skills and use them consistently will have an enormous positive impact on your projects. Poor team meetings are notorious for killing team cohesiveness and effectiveness. They kill time that people don't have to waste and end up killing people's enthusiasm for working together as a team. I know teams whose highest praise of their project manager is that "he runs good meetings." I know people who've been doing engineering and management for 40 years and still can't run a decent meeting. If you can, you'll be known by your managers and peers as someone who's mature, capable, effective, and unique.

Project management understanding and skills: Project management is not something one person bearing that title accomplishes alone. Projects are completed successfully because multiple team members know how to work together well, define a product and plan a project to create it, accomplish the designs, manage the risks, and move a product or service into production and delivery. Project managers look for people who take the initiative to do a good job both managing their own work and looking out for the team's overall success. Functional managers notice engineers who can manage work, especially in a cross-functional environment, and don't hesitate to give them more responsibility.

Communication skills: You realize that management needs certain information to do their jobs, as do other team members. You show respect for their needs by tailoring the information you provide. You help your department's work and your company's projects proceed with minimal surprises and problems by keeping people informed about aspects of your design that might affect them, open issues and risks, your current status, etc. Communication skills relate directly to good project management; your managers and other customers appreciate everything you do to make their work go more smoothly and help the entire endeavor be successful.

Presentation skills: Presentations are a particular form of communication that usually result in a very memorable impression of you with your audience. Boring, monotone, unfocused or unorganized presentations are all too common. Good presenters are prized. Not only do your customers appreciate it when you deliver what they need to know in an effective, interesting manner; management notices these skills when they are looking for mature, "presentable" engineers to send on customer visits and grow into positions of higher responsibility.

The above skills are directly applicable to your typical engineering and project work. Very significantly, they go way beyond the technical work we typically think of as the sole focus of the engineer's time. If done well as you simply execute your current job, these skills often automatically sell you to your customers for the next job. Believe me, you will stand out in the crowd. You will be sought out and fought over by functional managers and project managers alike.

If you don't currently possess these skills, your personal development plan must include learning them. (More on that in the next section.) You will use them throughout your career, no matter what position you're in. And as illustrated early in this section, now that you know why these skills are important to those who hire you or influence the jobs you will get, you can explicitly market your awareness of their importance and your personal capabilities as you look for your future opportunities.

One other skill is critical to your business development efforts: networking. You must be able to attend events, meet new people, and form connections and relationships based on common interests. And interestingly, your ability to network is actually a valuable skill to your customers. It shows that you're comfortable with new people; you're probably open to new ideas and able to learn from other people; and you're able to form relationships to get the job done—just like you'll have to in a new company or on a new cross-functional project team.

DEVELOPING YOUR SKILL SET

So, how do you develop all these skills? You have a number of options. You can take one skill at a time and develop it, such as by attending a class. Or, you can find an activity that helps you develop multiple skills at once, such as working on volunteer projects in your community or in a professional organization. For example, here are the skills I feel I've developed or enhanced through practice while doing IEEE volunteer work:

- *Business and marketing understanding*: financial and budget management; how to sponsor a booth at a tradeshow and gain sales leads from it; strategic planning and how business goals are linked with operational projects.
- *Meeting management*: practice running different types of meetings with a wide variety of personalities; exposure to new group decision-making tools; how to use project management techniques effectively for small projects.
- *Project leadership and management skills*: becoming a leader instead of doing everything yourself; practice motivating people when you aren't their primary project; how to overcome project delays due to distributed team environments—all your team members being scattered around the world.
- *Communication skills*: different forms of status reporting for different levels of executives and committees.
- *Presentation skills*: lots of practice making presentations—in a relatively non-threatening environment!

So do consider volunteer work in professional societies as one possible career development opportunity. As shown above, you'll gain skills in project management, budget management, dealing with all kinds of people, and so forth. And you'll meet an incredible array of people at different levels of various organizations. For instance, people I work with in IEEE are or have been senior managers or executives in well-known companies. You will learn a great deal; you may find a new job through these connections; you may even find a valuable mentor.

What are some other options for developing these skills? A number of ideas are outlined below; of course, you can come up with your own. In general, your "learning program" should include taking formal classes or seminars, tapping the experience of others, and finding opportunities to practice.

Business Understanding

- Take on some extra work on your project—offer to help Marketing develop the requirements specification, or just be very proactive about understanding the business justifications for your project and specific features within the product.
- Some companies allow functional rotation among groups—for instance, you can ask to work in Marketing for one project.
- Ask an executive to give a talk over lunch to explain aspects of the company's business, markets, contracts, etc., to your project team or department.

Broad Technical Expertise

- Pay close attention to the cross-functional specifications and issues on your projects. What will make your product manufacturable? serviceable? easily usable? Look at current products in the field and understand what cross-functional issues were not well addressed. Are customers complaining about the product? Is manufacturing having trouble producing it in volume? How can your current project do a better job? What suggestions might you make to the project manager and team to make this a reality?
- Attend a course on these subjects—university continuing education divisions routinely offer courses on design for manufacturability and design for usability.

Mature, Constructive, Flexible Attitude

- Self-examination is probably the most important first step here. Are you lost in a Dilbert mindset, cynical about management and proud of it? If so, you're unlikely to get the opportunities you want no matter what other skills you possess. Managers have too much to worry about already to want to work with someone who doubts managers' worth every minute of the day. The more proper attitude is that every company and project has management issues that can be just as complex as technical issues to solve—and you can probably contribute to their resolution.

To give an example: A particular project team is having a difficult time getting through the early project definition phase. Marketing wants a set of features that Engineering believes will be impossible to produce in the tight schedule timeframe required. Marketing won't budge. One engineer complains, disrupts team meetings, criticizes Marketing, and stirs up discontent and frustration. Another team member is more mature. He looks for outside help from another expert in the company to suggest possible tradeoffs. He encourages the troublesome team member to contribute his ideas for tradeoffs. He helps the somewhat shy project leader get these warring team members together to work out a solution together. He doesn't get upset when the trade-off decisions result in his favorite design challenge getting left until the next release of the product. This team member exhibits a mature, constructive, flexible attitude.

- What company, department, or project issues do you see that need work? How can you help their resolution, and encourage other team members to contribute as well? How can you exhibit this mature attitude especially during challenging times? You will most definitely be remembered for your demeanor and your contributions.

Meeting Management

- Read up on meeting management. This is one subject where a book can provide you with a great jump-start.
- Diagnose the problems you see in all the meetings you attend, and think about how you would solve them.
- Get some practice. Run a few meetings on your project, such as design reviews or educational briefings to cross-functional groups. Consciously practice using good planning and meeting management techniques in every one. Help any meetings in your volunteer organizations run better; offer to facilitate them or participate in their planning.
- Attend a meeting management class. Look for ones where you'll get a chance to practice and get feedback on your skills.

Project Management

- Take an introductory class to understand the basics of project management—project scope determination, planning, scheduling, tracking progress, managing risk, etc. Take more advanced classes in the areas of most use on your projects.
- Meet with your project manager and ask to understand their specific concerns on the project and how you can contribute to solving them.
- If your company does project "post-mortem" meetings or "lessons learned" meetings where teams analyze the project that just finished, find those reports and

read them. Identify what good and bad project management looks like in your company. Find ways to apply this new understanding on your current project.

- Consciously look for opportunities to put all these skills to use. For example, when you create your schedules for the next project, how can you make sure you've estimated your own work well? What risks do you see and what contingency plans should the project schedule include?

Communication Skills

- Look for examples of different types of project status reports in your company or ask for examples from your peers in other companies.
- Attend project status reviews to hear the level of information project managers present to executives. Observe closely to see how the executives respond to each presenter. Who do they seem to trust? Who do they "grill"? Which presentations seem to be short and sweet yet satisfactory to everyone? What questions do the executives ask?
- Ask your manager what you can do to improve your day-to-day communication. Then practice.

Presentation Skills

- Find venues for making presentations at technical or community organizations to which you belong.
- Offer to give a lunchtime seminar in your company on your area of technical expertise.
- Join organizations, such as Toastmasters, that are dedicated to developing public speaking skills.
- Attend a presentation skills class, especially one that offers the opportunity to develop a presentation, be videotaped, and get specific, personalized feedback.
- Approach someone in your company who is a good presenter and ask for feedback and coaching.

Networking

- Read a book on networking to get some simple tools for making yourself more comfortable in these situations, guidelines for following up with new contacts, etc.
- Practice, practice, practice. Find technical or professional meetings to attend whose subjects interest you. Vow to meet at least two new people at each meeting, and really connect with them. Practice the follow-up techniques.

Here is one last very important point: before launching off to do all of the above, sit down and think through a strategy for your learning program:

1. Identify what near-term and longer-term career options are most interesting to you.
2. Determine the skills you will need and succeed at them. Get feedback from peers, managers, and mentors.
3. Assess which of these skills you possess and which you might need to develop or enhance. Again, get feedback.
4. Plan the activities you should undertake to develop those skills. Prioritize the activities and create a timeline.

HOW TO MARKET YOURSELF CONTINUALLY AND DEVELOP FUTURE "BUSINESS"

Now we've looked inward a bit and talked about how to build a continual learning program for your ongoing career development. How do you look outward to market yourself and create opportunities?

Perform well. If you don't meet your commitments on a project, you won't get more responsibility on the next. Enough said. (This doesn't mean you aren't allowed to make mistakes. It simply means you must be viewed as a capable and committed member of your current department or team.)

Be known as someone who truly cares. Remember what we said about solving your customer's problems. The more you understand the challenges your project manager is facing, the constraints your cross-functional team members are operating under, and the design complexity your functional manager is dealing with, the more able you are to find ways to help. The more you help, the more respected and trusted you will be. Sincerely look for ways to serve your current and future customers.

Volunteer! Here again, volunteering is an excellent way to build marketing and business development into your daily life. Volunteer in the community, for your professional associations, and for in-company task forces. As you do this volunteer work, you market yourself indirectly (and as we said before, accomplish networking and skills development as well). The incredible advantage to volunteer work is that you advertise your worth by "doing"—you demonstrate what you're capable of and gain credibility. You earn your potential customers' *trust.* Effective marketing is really more about trust—created through demonstrated performance and relationships built over time—than it is fancy words, resumes, or brochures. I generally don't hire a person unless I've worked with them first—often in volunteer settings. As a personal testament to the business value of volunteering, I can draw a tree that shows how 80 percent of my consulting work in the last six years has resulted from three initial contacts at IEEE seminars that I helped coordinate as a volunteer.

Understand where your future customers are and find opportunities to interact with them. Do you want to move up in management? Volunteer for task forces where managers and executives will be present. Do you want to move into project management? Ask to participate in some cross-functional work on your project, such as helping with technical training of the sales force or working with the project manager to put together the overall schedule. Do you want to go into technical consulting? Volunteer for a professional society that offers related seminars, to meet the customers of this expertise and find out what they need. Join a consultants organization to learn from others. And remember, the more you get to demonstrate your wide variety of skills in these situations, the more you'll stand out and market yourself simply by doing real and valuable work.

Look ahead and position yourself for a higher-level job. If you want to take on increasing responsibility, whether as a technical contributor or as a manager, start your positioning for these roles "immediately." I am a firm believer in leaving all your future options open—by preparing consciously for those options and by not *precluding* any by even your earliest behavior in a company. (Remember the mature attitude—act mature and trustworthy from day one.)

How can you pave your way to a management position? Examine each of your assignments for what you can learn about management. What are the inherent technical management issues? What are the business issues? What is your manager concerned about on this project and how can you contribute to that manager's success? As we said before, demonstrated business understanding will make you stand out from the crowd. Those who are seen to understand business and management issues rather than sit around whining about deadlines and ambiguous situations are much more likely to get selected for management. Identify the roles you might be interested in; let people know you're interested; find out what specialized management or technical knowledge you might need to be considered for the job; and make this part of your development plan.

Understand what jobs/positions are available down the road: What should you do if you don't know what more responsible jobs are available, especially in the upper levels? Look for a receptive mentor or contact who would know. Look for task forces in the company that would give you more exposure to the upper levels of the company, and be willing to put in some extra time working on them. Some jobs get "filled" long before any opening becomes obvious. You've got to find some way to be "in the loop" and known as these positions open up. A side benefit to such involvement is that these task forces often result in new positions being created. You might get first pick!

Ask for the job. I got some great experience installing a huge image processing system at the customer site by asking to be on the installation and checkout team. I got my first management job as a group leader of hardware engineers by asking for it at the end of a project, well before the manager had come to any decisions about who to put in the role. Identify the job you might want to try, and don't be afraid to ask for it. Even if you don't get every one, you've demonstrated initiative and interest that will serve you well over time.

Be patient and nurture your relationships and contacts as you develop your skills.

Early in your career it's common to feel somewhat left out of the power structure. Realize that your day will come. As an engineer who had just moved into project management, I distinctly remember a conversation with a close colleague I'd worked with for a while. He had just gotten promoted to run part of the manufacturing organization. It suddenly occurred to me that if I was just patient, my peers and I could one day be "the ones in charge." Perform well now, do all the preparation we've talked about, support your colleagues in their careers, and be ready for that day. . .

Look for jump-start opportunities. Consider ways to jump-start your level of responsibility and visibility and track record. Sometimes the growth path in your company may be somewhat slow. Organizations truly have flattened. There are fewer traditional functional management positions and rungs up the ladder. You might want to join a start-up company to get access to wider and more cross-functional responsibility. My second job was at a start-up and it changed my life. I did technical work I had never had a chance to do as an engineer in a large company, became the director of an entire engineering department, and helped create a company from scratch.

CONCLUSION

My summarized recommendations for your career development come down to this:

- Understand your customers—how can you help them, be an excellent team member, and be the one they want to hire and work with for the long term?
- Identify the career paths you might want to take—what skills and experience will they require?
- Make a plan to hone your skills over time. Work to stand out now, demonstrating the technical and non-technical capabilities I discussed in this article, and prepare for future opportunities, too. Become special—"irreplaceable."
- Finally, give people the chance to see your skills in action, form long-term relationships built on performance and trust, and actively seek out the next opportunity.

With these steps, you've created your personal marketing "product plan" and product communications plan. You've set the stage for natural, ongoing personal "business development." Follow through, and you'll be set for a rewarding, long-lasting career.

ABOUT THE AUTHOR

Cinda Voegtli, BSEE, is president of Emprend LLC, a project management consulting and publishing firm in Silicon Valley. She has 17 years of experience in hardware and software development, engineering and project management, and product development process improvement in a wide variety of industries, including data and telecommunications systems, medical devices, database products, robotics systems, and virtual reality and game products. Before founding her consulting firm she held director-level engineering management and senior project management positions at several high-technology product development companies in California and Texas. She currently serves as president of the IEEE Engineering Management Society, and was guest editor of the winter 1996 issue of the *IEEE Engineering Management Review*, which focused on project management. She may be contacted at c.voegtli@ieee.org or at (650) 966-1650.

Lessons in Career Management from Silicon Valley: Key Factors for Continuous "Employment"

C. VOEGTLI

ABSTRACT

The market for technical and management consultants is booming. While not everyone is interested in that work model, it is becoming an increasingly important segment of the available career opportunities for engineers—one that should continue to grow.

Successful consultants exhibit important attitudes, approaches, and skills that can be learned by engineers who want to become consultants. Engineers within the corporate world can apply these attributes to maximize personal opportunities in the face of the changes and uncertainties created by downsizing, restructuring, corporate alliances, and market dynamics. These important areas include past and current job performance, relationships, marketing and networking abilities, people skills, personal presentation, flexibility, and continuous learning abilities.

This paper provides a brief overview of the current state of the consulting arena for electrical and electronics engineers, including influences on the availability of consulting jobs, where the jobs are likely to be found, and the functions for which consultants are typically hired. It then provides examples of specific successes and failures of engineers who have moved into technical and engineering management consulting in Silicon Valley, based on the presence or absence of the crucial abilities mentioned above. It also provides examples of corresponding successful career transitions within the corporate world based on these competencies. From these examples the paper summarizes the critical lessons to be learned and the specific techniques used in these areas of competency, which can be applied by the engineer seeking to widen career options and stability, whether inside or outside the traditional corporate environment. Finally, the paper provides guidance for developing or improving the necessary attitudes, approaches, and skills.

Cinda Voegtli, President, IEEE Engineering Management Society, and President, Emprend LLC

THE CURRENT OUTLOOK FOR CONSULTANTS

Impacts on Job Availability

Downsizing. Downsizing is a positive influence in that companies are now highly conscious of keeping overhead costs low. The incessant media coverage of downsizing and future employment trends has brought attention to the option of using consultants. Companies will consider consultants for a number of project- or support-related jobs. The "movie production" analogy, where people come together for a particular project, then disband, is now more accepted in the high-tech product development world. The only downside to this trend for aspiring consultants is that more people are competing for the available jobs. Fortunately, as this paper will explain, a consultant can differentiate himself adequately to be the one who is hired.

Time-to-Market concerns: Heightened emphasis on reducing the time-to-market for new products means that speed is often more important than development cost, leading to the decision to hire extra manpower to accelerate project completion or to mitigate technical risks by using technical consultants to develop alternative designs.

Technological advances: The rapid pace of technological innovation has created numerous technical "niches"; when a company needs an expert in a technology, it will often turn to a consultant.

Types of Jobs Available

The types of jobs readily available to consultants fall into three categories:

Engineering work, including integrated circuit design, digital and analog electronics design, and software development, compliance (EMI), software, product engineering, and testing.

Related technical functions, such as design verification and quality assurance testing, technical writing, product engineering, regulatory certification, network administration, and computer-aided design.

Related management functions, such as interim staffing for vacant management positions, project management contracts, and product development process improvement consulting.

Where the Jobs Are

Large companies: Large corporations are most worried about "head count" when cost controls are necessary. Hiring consultants can be more attractive because the payments to a consultant are an expense, not a fixed labor charge, and because the corporation does not have to pay for such overhead items as benefits and facilities. The use of consultants is also "lower risk" for the companies; if the project is canceled, the consultant's contract can simply be terminated.

"Medium"/merging/lean companies: These companies may be just starting to launch simultaneous projects. They have not staffed up yet to handle multiple projects while sustaining work for their existing products, so they have incredible manpower needs. These companies also usually do not have the infrastructure of larger companies, meaning they are more likely to have openings for support positions, such as computer network administrators or software test engineers. These companies are eager to find people who know the latest methodologies and tools.

Small, "bare-bones" companies, including start-ups: Here the head count depends on the company stage and funding. Many start-ups are not glamorous, highly funded companies. They are under incredible pressure to get their first product to market as quickly as possible, yet since they are in a risky stage in an undercapitalized state, they may not want to risk staffing up. They may also have the problems related to lack of infrastructure. Some may have one product in the field whose sustaining work is consuming all technical resources. Such a company may use consultants to help get started on the next product.

SUCCESSES AND FAILURES IN THE CONSULTING WORLD

The following true stories from my direct experience in the last five years illustrate some of the factors that influence a consultant's success.

Paul was an engineering manager in a computer company. He was both managing and personally doing integrated circuit logic design. His company went through some direction changes, the next project was canceled, and he decided to leave. Paul wanted to do IC logic design for other companies, but he discovered that most of these companies expected designers to know a particular design and verification tool. Paul did not have experience with the tool, so he took a local class that afforded both theory and hands-on use on a project. He subsequently landed his first contract job as an IC designer and has completed five 8- to 15-month-long projects since, with some down time in between, which he used to look for the next contract.

Jim was an engineer who had worked as a manager and vice president of engineering in a small company. He left the company after an acquisition and tried to hire himself out as a management expert who could advise product development projects. In his own words, no one wanted him. He then compiled all his management wisdom into a course on product development and offered it through the local IEEE chapter. He offered it once or twice a year over the next three years. Because the IEEE chapter kept inviting him back, other people took notice and asked him to speak at local conferences. Eventually, all these speaking gigs resulted in on-site company training and consulting. His company is now successful enough that he has hired several more consultants. He had to live with less income for several years, but his income potential is now very high.

Six years ago I left a corporate project management position to work as a lone consultant, managing client projects on a contract basis. In my first three years in business I actively looked for an engagement. I had steady project management contract work based on referrals from people with whom I had worked previously. For example, my first consulting job was to train my replacement at the corporation I left. My second job was to write a technical manual for a start-up run by friends of the last corporate vice president for whom I had worked. Note that my intent was to do project management consulting, not technical writing! However, this company needed the manual first and I wanted to get in the door, so I did the manual. I then got a second contract to manage the certification process for their new product line. My next client came from someone I spent 10 minutes talking to at an IEEE seminar; I had three successive contracts with that client. My subsequent job came from a referral by one of the consultants I worked with there. Over time I decided I wanted to share my experiences in a workshop setting to reach more people, and connected with the person with the consulting/workshop company mentioned above. Now my company does project management-related workshops, consulting and project team coaching, product development process consulting, and publishing. Each type of work or product feeds the other. And it was all built from a few initial relationships and continued networking.

Within the corporate arena, Karen demonstrated an entrepreneurial style in progressing through an electronics design career. After 10 years as a logic (board) designer, she decided she wanted to get into ASIC design. It appeared to be difficult to get a company to hire her at her current salary level, since she had no ASIC design experience. She persevered, and eventually found a well-known company that was willing to take a chance. After a successful project, she saw a trade publication advertisement for another well-known computer company for which she had always wanted to work. The particular project was not mainstream for the company and was not exactly what she wanted in the long run, but she wanted to work for this company. On the strength of her resume and interview she landed that job as well. After two years that company was acquired by an even larger, better known Silicon Valley company and she was able to transfer to a new computer IC design project. On the strength of her performance and relationships there, when the next round of projects fell apart she had interviews lined up within days, got lucrative offers from several companies, and went on to a start-up (a long-held career dream).

Note that the consulting model chosen can have an influence on success and failures. Hector created an outsource product development company, meaning he would do development work for other companies. Later he wanted to provide product development process consulting. His approach was to try to sell high-powered consulting to clients for whom he was already doing product development work. Unfortunately, he was always too busy managing the development projects to get the other side of the business launched quickly. Unlike Jim, who abandoned his "high-powered management expert" selling approach and infiltrated companies through his seminars, Hector's company is still struggling to establish that business.

There are actually two categories of "failures" in the consulting business. In the first, the consultant never finds enough work to support himself adequately. In the other, the consultant fails to get invited back to a previous client, or as one colleague put it, does not make it into the employer's Rolodex. In this case the consultant is either not as successful as he could have been, or the consultant eventually fails altogether because of the difficulty of finding new clients over time.

For example, Alan tried very hard to build a consulting business. He had a wide range of skills related to computer systems, hardware and software development, and project management. He tried to market all of these skills accordingly to find whatever type of job was available. Unfortunately, he had several problems. Some people thought that he could not possibly be that good in so many different areas, so they did not believe his resume. Additionally, he picked work opportunities that required bidding processes (developing proposals and waiting for notification of winning or losing). This is a very slow, risky way to get business as a lone consultant. Finally, his diction, weight, and dress combined to make a less-than-professional impression on some people. He did not inspire confidence. He was not able to land enough consulting work, so he eventually had to augment his income from other non-engineering sources.

As an example of the second mode of failure, Kevin is a software engineer who, after a number of years in a corporate environment, launched a solo consulting business. He provided software engineering services to various companies. He came across as a knowledgeable engineer; his personal presentation was such that hiring managers felt comfortable dealing with him; his experience list was impressive enough for him to get invited in for interviews solely based on his resume. He seemed to have enough business to pay his bills. However, he did not make it into my Rolodex. In working with him, I felt he never wanted to solve the detailed problems on my project; he wanted to "advise" me but not take responsibility and get his hands dirty. Therefore, he did not help me very much. In addition, this person did not win over the vice president of the division for the long term. The consultant had taken the first project at that client without estimating the complexity of the product to be developed and the resulting workload accurately. When the implementation proved difficult, the schedule slipped, and the resulting system was below expectations. The perception was that the consultant screwed up the project. He was not invited back. Note: the proper setting of expectations is a concern for engineers within corporations as well. Your performance is rated based upon expectations. For a consultant, perceived performance simply becomes critical more quickly because the client can let you go at any time. They do not wait until your next performance review period to assess you!

In another case, Steven's failure as a consultant resulted from issues of perceived performance and personal presentation. Steven worked for a consulting firm. He was not an incredibly fast worker. He seemed to take too long to "catch on" in new work situations. In addition, his personal presentation did not inspire confidence. During stressful project situations he was very open about the difficulties in dealing with the situation. To be sure, during stressful projects a team tends to bond together, and commiserating about a tough project situation is common. However, it is not a good idea for a consultant to let down his guard and participate too actively. The consultant is constantly being evaluated as to his worthiness, and keeping the employer's confidence is very important, especially in the stressful situations for which consultants are often hired. This does not suggest that a consultant should lie about the state of a project. However, one *should* always come across as competent. Steven's perceived capability and performance did not rate a return engagement with this client. This situation was repeated at several more clients of the consulting firm. Since his performance did not get the consulting company "invited back," there was eventually not enough work to pay him sufficiently. He left to take a full-time job.

THE IMPORTANT CAPABILITIES

These success and failure examples lead to a summary of the capabilities I believe are important to the success of an aspiring consultant. They are also important for keeping your job in today's corporate environment and for finding new opportunities in the corporate world.

Performance (past and current): Your past performance often gets you into an employee position. It is even more critical to finding work as a consultant. Be able to prove your past performance. Have a resume that shows the successful completion of a product. Keep a list of satisfied clients with letters of reference, if possible.

Your current performance on your job or contract will determine how long that current job will last and whether the next one will be forthcoming. The key to successful consulting is getting repeat business from happy clients.

Relationships. Relationships are everything. People hire those they trust or those a friend trusts. Every job or every contract must be looked at as a chance to form relationships. You must form good ones with the people you work with, those most in the position to have an opinion about your performance and your desirability for the next job. Your client must want to use you for the next contract. In the corporation, your peers and managers must want to use you on the next important project. For example, in large companies groups band for a project, and then move on to others. Where you get to go next will depend on your past performance, and who knows about your past performance!

Networking abilities. You must continue to meet new people to maximize your work potential. In Silicon Valley people move around often, creating a potentially far-reaching web of contacts. Even in environments where moving is not so prevalent, keeping this perspective can help create more opportunities. Your networking efforts must be on-going and should include:

- Keeping in touch with former associates, both peers and managers (in all functions, not just engineering) and with other consultants.
- Attending related organization meetings: technical or management-related (such as IEEE Engineering Management Society or the Institute of Management Consultants), user groups for development tools, and entrepreneurial groups.
- Volunteering with organizations in both industry and your community. Three of my long-term clients resulted from contacts I made working as an officer in the IEEE Engineering Management Society, helping put on monthly meetings and seminars. Volunteering in community organizations is often the best way to meet some important figures who could provide opportunities later: you generally won't run into large-company CEOs at IEEE functions, but you might in a community setting.
- Attending continuing education classes (such as quality assurance or project management). Examples are university extension courses, IEEE workshops, and vendor tools classes.
- Visiting conferences and trade shows and meeting other participants (including influential ones) within your industry.

Marketing and sales. Aspects of marketing such as direct mail apply more to growing product-based businesses. I will not discuss those here; as an individual consultant, a large part of marketing is conducted through the relationships and networking discussed above. In the sales arena, here are important attitudes and techniques for selling yourself and for differentiating yourself from potential competition:

- If you are competing for a contract or job opening, find out the hiring manager's name and call him or her directly to call attention to your candidacy.
- Always emphasize your interest in helping solve the manager's or company's problems.
- Be ready to address any company concerns about hiring consultants, which include IRS status, your ability to do the job, or your rates. Follow the guidelines for being a legitimate independent consultant and be able to show the documentation. Or go further and incorporate yourself so that there is no barrier to hiring you. (Incorporating also says that you're serious about being in business.) Have a great resume, summary of past consulting engagements, and letters of recommendation from past clients. Develop various payment structures so that you can bid based on the client's needs and concerns.
- Be able to recommend other qualified consultants when you are not right for the job. Being honest and helping solve their need will get you into the Rolodex for the next opportunity.
- Discuss your commitment to training the client on your work product so that they perceive long-term benefit from hiring you.

Personal presentation. People draw conclusions, correctly or not, from your appearance. They draw conclusions about your importance, competence, and success from the way you dress, speak, write, carry yourself, and interact with others. You must pay attention to the culture and expectations of your potential clients so that you do not unwittingly lose business due to personal presentation issues. Dress appropriately for the situation, be well-groomed, practice speaking with new people, get your resume and other written "sales tools" reviewed by someone else.

People skills. Are you comfortable talking to new people? Do people feel they can work with you? Can you handle working with people who are not like you? Consultants constantly plunge into new situations and must deal with new groups of people on each contract. Just as potential employers want to be sure someone will fit in, clients want to be sure that the consultant will not only perform his part of the project well but also be an effective team player. Note that these interpersonal skills contribute to both the relationship-building and networking discussed above.

Flexibility: Often it may be advantageous to get in the door at a particular client, even if the first work they have available is not exactly what you had in mind. (But a caution: You must do a good job at this work! If the client perceives that you are not doing excellent work because you are not really interested in it, this approach will backfire.) Clients may also value your ability to be flexible in the kinds of work you can perform. This holds true especially with smaller companies that need a jack-of-all-trades. Some suggestions:

Be multi-skilled. Be knowledgeable about tools; have both hardware and software skills; know how to write good specifications and other product documentation; understand the entire project cycle. (This advice applies to both engineering and management consultants.)

If you feel you need experience in a particular area, consider "gratis" or reduced-rate work if it is the only way to gain that extra experience. You could subcontract from another consultant or help a start-up that can't pay much (or anything) yet. Some continuing education classes call for intensive projects writing software or test plans or marketing plans. If the subject of the project is chosen carefully and executed well, it can provide legitimate experience.

A final way to gain extra experience is through volunteer work. For instance, when I first became active in IEEE I volunteered to help organize their annual seminars. At the time I had no intention of ever teaching seminars myself. It eventually became part of my business, however, and the lessons I learned helping IEEE made my transition to this new business much smoother.

As an employee, you should understand whether your company, project, and/or boss value such flexibility and act accordingly. Flexibility is valued in corporations after a downsizing, when the remaining people are asked to do a wider variety of tasks. Flexibility can also be a great way to get visibility or "extra points" in corporations. In some Silicon Valley companies, a criteria for receiving an excellent performance review is to contribute significantly to the project *outside* the scope of your normal job description.

Continuous learning. Technology and management practices continually evolve. As an employee or a consultant you must continue to learn new skills and perspectives to be employable. Consultants have to worry more constantly about marketability; continuous learning is critical for both differentiation with respect to competitors, and for the above flexibility in work.

HOW TO DEVELOP THESE SKILLS

All of these attitudes, approaches, and skills can be developed. First of all, examine your past experiences honestly and try to pinpoint the reasons for your successes and failures in previous jobs and other endeavors. Ask other people as well. Look at your old performance reviews.

Then identify the areas that need work to increase your chances of success in the consulting or the corporate world, and formulate a plan for improvement. Include the following activities:

- Attend classes in the areas needing improvement. Local community colleges often have short, focused classes on personal development, speaking and writing skills, socializing, etc.
- Join organizations devoted to a particular area and participate actively. For help with speaking skills, try the Toastmasters program, which has a structured program for learning to speak confidently in front of other people on a wide variety of subjects.
- Look for mentors. A mentor can give you invaluable advice and save you a great deal of time. They can help you avoid common pitfalls by providing you specific warnings; they can give objective assessments of your strengths and weaknesses; they can point you to other resources. A mentor could be a previous boss or peer who has progressed further than you along a desired path, or a colleague from an organization with whom you have developed a relationship. A mentor could actually be an organization designed to provide peer-level guidance. The IEEE Consultants' Network provides such opportunities, as do a number of entrepreneurial-related groups.
- Watch successful people. Go to conferences and organization meetings and observe the speaker and other influential participants.
- Volunteer for new activities and get some practice.

CONCLUSION

The world of work has changed significantly, from the prevalence of life-long corporate employment to the reality of shorter-term corporate stints and less stable project-based work. But opportunities are abundant for doing exciting work and making an adequate or even sizable income. With on-going technology advances and the evolution of corporate organizational structures, the opportunities for consultants should continue to grow. I have identified the factors that from my experience in Silicon Valley are instrumental to the success of consultants. Keeping these factors in mind, with some initiative, hard work, and perseverance, engineers can make a successful career transition to the consulting world.

ABOUT THE AUTHOR

Cinda Voegtli is current president of the IEEE Engineering Management Society. She is president of Emprend LLC, a project management consulting and publishing firm in Silicon Valley, and is a consulting partner with Global Brain Inc., a consulting firm that focuses on techniques for rapid product development. Previously she held director-level engineering management and senior project management positions at several high-technology product development companies in California and Texas. Ms. Voegtli holds a bachelor's degree in electrical engineering. She can be reached by e-mail at c.voegtli@ieee.org or by telephone at (650) 966-1650.

Effects of Humor on Social Influence Strategies in a Workplace Scenario

M. WALTERS

ABSTRACT

Humor may be an important factor in the effectiveness of power strategies. This study investigates the effects of humor on the original power bases developed by French & Raven (1959). Reward, Coercion, Legitimate Dependent, Expert, Referent, and Legitimate Position powers were examined in both humorous and non-humorous conditions. Participants completed a survey, in which they rated the effectiveness of a supervisor using the power bases and humor conditions in a workplace scenario. The survey indicated that the use of humor had a negative effect in the Coercive, Expert, and Legitimate Dependent power conditions. This paper discusses implications for future research concerning the use of humor and social influence techniques.

INTRODUCTION

Little research has been conducted to examine the effects of humor on influencing behavior. These effects can be important in relation to management technique and employee satisfaction. There are contradictory results and inconclusive evidence from studies that have attempted to assess the relationship between the two variables, but few experiments have examined the relationship between French and Raven's (1959) bases of power and humor. This study tests for the effects of humor on specific influence techniques by looking at the effectiveness ratings of supervisors who use the bases of power: Reward, Coercive, Legitimate Dependent, Referent, Expert, and Legitimate Position, in humorous and non-humorous ways.

HUMOR

Our experiment will be examining the effects of humor on power. However, a prob-

Michael Walters, University of California, Los Angeles

lem arises when attempting to define humor operationally and to examine its position in the workplace.

Use of Humor in the Workplace

Many studies have attempted to determine the role of humor in office-type settings and have discovered the difficulties of testing for humor in real-life environments. An article by Ullian (1976) studied the prior results from experiments that focused on the functions and meanings of humor in organizations. He found that joking has been theorized to accomplish various tasks, such as to release tensions among employees (Beadney, 1957), affect role-sending in the organization (Coser, 1959), reduce boredom among workers (Roy, 1960), and help workers discover possible sexual partners from among the members of their work group (Sykes, 1968). However, studies could not show which theories contained significant merit.

An experiment by Ullian (1976) examined the context of joking in an organization. The experimenter looked at the context of individual jokes, the participants, time, rate of activity, location, and referent in a field study design. Ullian concluded that joking is not a random behavior but occurs in definite patterns. In other words, certain workers joked more than others, were joked with more than others, and were targets more than others. The implication of this finding is that humor has a specific role in the workplace, but it is extremely difficult to define humor's role operationally in every workplace.

Potential for Harm

Although there is some evidence to support the claim that humor has a role in the workplace, other evidence indicates that humor can have a negative effect on influence techniques in workplace scenarios. A study by Wimmer (1994) addresses this issue concerning factors involving the use of humor in mediation tactics. In general, four areas were discovered where humor was found to have a negative impact in mediation: timing, the mediator's personality, ethics, and power differences. First, inappropriate timing may disrupt the flow of a conversation or display potential awkwardness of the mediator in discussing a certain topic. Second, the personality of the mediator was found to have a potentially negative impact because not everyone is comfortable using humor and, thus, may come off as rigid or awkward. Third, the humor must adhere to certain ethical limits, or it may appear vulgar to a target. Fourth, power asymmetry was discussed as a limitation on the use of humor because if used improperly and to the detriment of the weaker party, such a technique can invalidate and discredit the whole concept of mediation. As a result, Wimmer's discussion, taken in combination with previous studies, indicates that there is a potential for the use of humor in power use settings, but care must be used in the way the humor is used in those particular settings.

POWER BASES

Definitions

French and Raven (1959) introduced the original six bases of power in an attempt to explain the various ways in which people influence others. French and Raven found six main techniques of influence, which they named Reward, Coercion, Legitimate, Expert, Referent, and Informational power. These were later differentiated further, and the concept of environmental manipulation was added (Raven, 1983). The influence techniques were defined as follows:

1. *Reward Power* stems from the agent's ability to grant some reward to the target.
2. *Coercive Power* involves threat of punishment by the agent toward the target.
3. *Legitimate Power* is based on social norms, such that the target feels an obligation to comply with the agent's requests.
4. *Expert Power* is based upon the trust that the target has in the superior knowledge, ability, and/or truthfulness of the influencing agent, rather than relying on the persuasiveness of the information or logical argument.
5. *Referent Power* is based upon the target's ability to identify with the agent.
6. *Informational Power* is similar to Expert power, except it is based on the information or logical argument that the agent can present to the target.

Raven (1992) further explained differences that involve the Legitimate power base. It appears this power base can be divided into four separate forms: Legitimacy of Position, Dependent, Equity, and Powerlessness. For the purposes of this study, we examined only the Legitimate Position and Dependent techniques of influence, which may be defined as the following:

1. *Legitimate Position Power* is based upon the belief that a target should or is obliged to conform to the agent's orders because the agent is, after all, the superior.
2. *Legitimate Dependent Power* is based upon the belief that the agent is depending and counting on the target to accomplish a task for him or her (Berkowitz & Daniels, 1963).

In our study we will use Reward, Coercion, Legitimate Dependent, Referent, Expert, and Legitimate Position to examine the effectiveness of a supervisor.

Effectiveness

In the current study we examined the effectiveness of supervisors who utilized the power bases. A study from Litman-Adizes, Raven, & Fontaine (1978) examined this exact relationship. They found that the most effective technique in inducing private acceptance of change from a supervisor was the Informational power. They found Information, Reward, and Referent power were most conducive to mutual evaluation and liking, while Coercive and Legitimate power were least effective in both respects. Litman-Adizes, et al. also found that participants were more likely to attribute compliance to the worker's will if Referent, Information, or Reward power was used and less so if Coercion or Legitimate power was used. The results from this study indicated that Information, Reward, and Referent power were rated as more effective influence techniques, while Coercion and Legitimate power were least effective in terms of influencing subordinates' ratings of effectiveness. The purpose of this study was to determine whether these results remain valid when examining the effects of humor.

HUMOR AND INFLUENCE

As stated previously, there have been many inconsistencies with studies examining the effects of humor on influencing others' behavior. Markiewicz (1974) discussed humor's effects on changing attitudes through persuasive messages. The author found by examining all previous relevant studies that (1) humor integral to or adjacent to a persuasive message does not influence persuasion significantly; (2) humor's effects on comprehension and source evaluations are inconsistent; and (3) retention does not appear to be altered by humor use. The results indicate that humor has no effect on persuasion, and the author points out that this may be due to severe methodological problems with experimental designs, inadequate control messages, and questionable humor manipulations. However, it is obvious that this study deals a blow to any hypothesis related to humor's positive effect on influence.

On the other hand, a more recent study by O'Quin & Aronoff (1981) that looked at negotiation and bargaining as influence techniques found opposite results from Markiewicz (1974). In this study participants received an influence attempt from a confederate that varied in size and was administered in either a humorous or a non-humorous way. The results indicated that the use of humor led to an increased financial concession (more willingness to submit to influence attempt), more positive evaluation of the task and marginally lessened self-reported tension, but did not increase liking for the partner. Thus, O'Quin & Aronoff demonstrated that humor had a positive effect on their influence attempt, which conflicts with Markiewicz's research.

HYPOTHESES

This study examined the effects of humor from a supervisor who was attempting to influence a subordinate to complete a task. Participants rated the effectiveness of the supervisor who used the six power bases in either a humorous or non-humorous manner. It is our belief that (1) Reward, Referent, and Legitimate Dependent users will be viewed as more effective supervisors in humorous conditions because of their friendlier and more positive manner of influence, and (2) Coercion, Expert, and Legitimate Position powers will be viewed as least effective supervisors in non-humorous conditions because of their harsher and more negative manners of influence. These hypotheses are based in part on the results from Litman-Adizes, et al. (1978), who showed that Information, Reward, and Referent power were rated as the more effective manners of achieving influence by a supervisor, while Coercion and Legitimate power were rated as least effective.

METHOD

Participants

Undergraduate students completing an introductory psychology requirement were used as participants in this study. In all, 81 students (41 men and 40 women) of mixed ethnicity were surveyed and treated in accordance with the "Ethical Principles of Psychologists and Code of Conduct" (American Psychological Association, 1992).

Materials and Procedure

A pen-and-paper survey was developed to examine the effects of humor on power bases. Participants were asked to rate a supervisor who had just made one of 12 statements. The 19-page survey contained sections describing the power base used by the supervisor in either a humorous or non-humorous manner, five questions designed to rate the effectiveness of the supervisor, and an adjective rating scale designed to test for the perceived humor rating of the supervisor (see Table 1).

At the beginning of each survey, participants were instructed to imagine themselves working on a job when their supervisor asks them to change what they are doing. They were told the supervisor had asked them to change before, but they had not done as he/she had asked. No mention was made concerning the specific gender, personality, or appearance of the supervisor. Participants were then presented with a statement by the supervisor using one of the six influence techniques designed to get them to change their behavior as he/she had asked. After reading this statement the participants were asked to rate a series of five questions on a seven-point scale, to assess the supervisor's effectiveness. Finally, participants were asked to rate the supervisor for humor on a seven-point adjective rating scale. These three main sections—influence technique, effectiveness questions, and humor ratings—were then repeated along all six power conditions. Participants read three conditions of influence techniques using humor and

TABLE 1. Mean effectiveness rating of supervisors by power condition and humor condition.

	Humor Condition			
	Humorous		Non-Humorous	
Power Condition	M	n	M	n
Coercion	3.55*	40	4.24*	41
Legitimate Position	3.55	40	3.37	41
Expert	3.76**	40	4.20**	41
Reward	5.00	40	5.16	40
Legitimate Dependent	4.24*	41	5.46*	40
Referent	4.44	41	4.24	39

*Significant at p < .05.
**Significant at p = .10.
Note: The higher the score, the more effective the supervisor was rated.

three without humor. Thus, every subject saw all six power conditions, half with humor and half without humor. The six power conditions were randomly distributed in each survey, to minimize the effects of order in the results.

Independent Variables

The independent variables in this experiment were the six power bases (Reward, Coercion, Legitimate Dependent, Referent, Expert, and Legitimate Position) and the humor condition (humorous and non-humorous). The statements were developed after extensive pilot-testing in an attempt to control for the rating of humor of the individual statements. Each participant received the following influence statements, with humorous additions made simply at the end of the influence statement.

1. *Coercive Statement:* "It is really important that you do this. Unless you do what I ask this time, I shall have to put something in your personnel file saying that you have neglected to do what I asked."

   Humorous Addition: (As an afterthought, laughing) "I also will remember to report that awful tie you wore to work the other day."

2. *Legitimate Position Statement:* "It is really important that you do this. As you know, after all, I am your supervisor and you should feel obliged to do as I ask."

   Humorous Addition: (As an afterthought, laughing) "I guess it's like being the chief Power Ranger, so I guess you know what I mean."

3. *Expert Statement:* "It is really important that you do this. I can't at this moment explain to you exactly why, but believe me there are good reasons. You will have to trust me, since I really have information about this which you don't have."

   Humorous Addition: (As an afterthought, laughing) "What reasons? I'm operating on top secret information from the Pentagon."

4. *Reward Statement:* "It is really important that you do this. If you do as I ask, I shall put in a good word for you in your file which will be helpful to you in the future."

   Humorous Addition: (As an afterthought, laughing) "Not only would you get the good word in your file, but you will also receive a free lollipop at the company's expense."

5. *Legitimate Dependent Statement:* "It is really important that you do this. I am really depending on you taking care of this for me, since I cannot finish my task until you complete this one."

   Humorous Addition: (As an afterthought, laughing) "This isn't to put any pressure on you. I mean don't let the fact that I just bought a new house and have three small children to provide for pressure you into working any harder than you already are."

6. *Referent Statement:* "You know we have been working together on this job for a while and we really have a lot in common, don't we? I would think that we would see eye-to-eye on things like these and that you would see things as I do . . . So how about doing what I ask?"

   Humorous Addition: (As an afterthought, laughing) "I don't mean that we're two peas in a pod . . . that would have to be a pretty big pod wouldn't it . . . but you know what I mean."

Dependent Variables

The dependent variables for this experiment were measured on an effectiveness scale. The results were scored on a seven point scale, where a score of 1 indicated Completely Ineffective and 7 indicated Effective to a Very Great Extent. The effectiveness scale was composed of the following questions: (1) How effective would you feel this supervisor was (as a supervisor)?; (2) What is the likelihood that you would do exactly as your supervisor asked?; (3) How much would you like or dislike the supervisor personally?; (4) To what extent would you have complied because that was what the supervisor wanted you to do?; and (5) To what extent did you change your behavior because that was what you really wanted to do?

In an attempt to control for the perceived use of humor in the humorous conditions, an adjective rating scale was used to measure how funny participants perceived the supervisors to be. The results were measured on a three-item humor scale, which participants were asked to measure, along with 12 other distracter items. The items in the scale were dull/witty, non-humorous/humorous, and serious/funny, and participants rated them on a seven-point rating scale.

RESULTS

To determine whether the means between the individual power conditions of humor and non-humor were statistically significant, several *t*-tests were performed for the effectiveness scale and humor ratings. In general, the significant findings indicated that the non-humorous conditions were perceived as more effective than the humorous ones in the Coercive, Expert, and Legitimate Dependent conditions. The humor statements were all rated as funnier than the non-humorous statements. However, the scores were low on the humor rating.

Humor Ratings

All humorous conditions were found to be rated as funnier than the nonhumorous conditions. However, the mean scores of the humorous ratings were very low, where a score of 1 indicated not funny and 7 indicated very funny (see Table 2). An alpha level of .05 was used for all statistical tests. For the Coercive statement, the humorous condition ($M = 2.14$, $SD = 1.09$) was rated as funnier than the non-humorous condition ($M = 1.40$, $SD = 1.28$), $t(77) = 2.82$, $p = .006$. For the Legitimate Position statement, the humorous condition ($M = 2.62$, $SD = 1.35$) was rated as funnier than the non-humorous condition ($M = 1.33$, $SD = 1.05$), $t(73) = 4.78$, $p = .0001$. For the Expert statement, the humorous condition ($M = 2.93$, $SD = 1.50$) was rated as funnier than the non-humorous condition ($M = 2.07$, $SD = .96$), $t(66) = 3.03$, $p = .004$. For the Reward statement, the humorous condition ($M = 4.00$, $SD = 1.20$) was rated as funnier than the non-humorous condition ($M = 2.65$, $SD = 0.69$), $t(64) = 6.21$, $p = .0001$. For the Legitimate Dependent statement, the humorous condition ($M = 2.80$, $SD = 1.19$) was rated as funnier than the non-humorous condition ($M = 2.3$, $SD = .90$), $t(74) = 2.16$, $p = .03$. For the Referent statement, the humorous condition ($M = 3.42$, $SD = 1.16$) was rated as funnier than the non-humorous condition ($M = 2.20$, $SD = .89$), $t(74)$, $p = .0001$.

Effectiveness Scales

For the Coercion condition, participants saw the non-humorous statement ($M = 4.24$,

TABLE 2. Mean humor rating of supervisors by power condition and humor condition.

Power Condition	Humor Condition			
	Humorous		Non-Humorous	
	M	n	M	n
Coercion	2.14	40	1.40	41
Legitimate Position	2.62	40	1.33	41
Expert	2.93	40	2.07	41
Reward	4.00	40	2.65	40
Legitimate Dependent	2.80	41	2.65	40
Referent	3.42	41	2.20	39

Note: The higher the score, the funnier the supervisor was rated.

$SD = 1.31$) as more effective than the humorous statement ($M = 3.55$, $SD = 1.30$), $t(78) =$ 2.37, $p < .05$. In the Expert condition, participants viewed the non-humorous statement ($M = 4.20$, $SD = 1.08$) as marginally more effective than the humorous statement ($M = 3.76$, $SD = 1.29$), $t(76) = 1.65$, $p = .10$. Similarly in the Legitimate Dependent condition, people perceived the nonhumorous statement ($M = 5.46$, $SD = 0.75$) as more effective than the non-humorous condition ($M = 4.24$, $SD = 1.10$), $t(70) = 5.83$, $p < .05$.

The results from the Reward, Referent, and Legitimate Dependent power conditions showed insignificant results. The effect of humor on Reward power was not statistically significant, $t(71) = 0.55$, $p > .10$. The effect of humor on Referent power was not statistically significant, $t(76) = -0.83$, $p > .10$. The effect of humor on Legitimate Position power was not statistically significant, $t(78) = -0.66$, $p > .10$.

DISCUSSION

The results of this experiment, which tested whether humor has a positive or negative effect on the Reward, Coercion, Legitimate Position, Referent, Expert, and Legitimate Dependent power bases, have shown that parts of the original hypotheses were correct and parts were incorrect. These differences may be attributable to problems in the experiment, but shed light on corrections that may be made in the design for future testing on this subject.

Major Findings

It was originally speculated that Coercion, Expert, and Legitimate Position powers would be considered more effective in a non-humorous condition, and it was shown that the non-humorous Coercive, Expert, and Legitimate Dependent conditions were more effective. The indications that the effect of humorous influence techniques were negative supports earlier research done by Markiewicz (1974). This effect was expected in the case of Coercive and Expert power because the use of humor might harm the effectiveness of the supervisor when he or she uses threats of punishment for Coercive or trust in the experience and wisdom of the supervisor in order to influence the worker for Expert power. The use of humor might undermine the legitimate authority of the supervisor because in attempting to use humor in the Coercive and Expert influence techniques the supervisor might be giving him/herself the appearance of being hypocritical. In other words, if a supervisor threatens a subordinate to change his or her behavior and then cracks a joke about what he or she just said, a subordinate might feel confused about the intended meaning of that threat and about that supervisor's effectiveness. In addition, if a supervisor portrays him or herself as one who has superior knowledge of an experience over a subordinate and then tells a joke about the situation, a subordinate might feel confused about the merit of the supervisor's expertise or seriousness of the situation.

The non-humorous Legitimate Dependent condition was found more effective than the humorous one as well. This may be caused by perceived seriousness of the participants related to the importance of a supervisor relying on them to complete a given project. This influence technique implies there is a bond being established between worker and supervisor, and the use of humor in an important dialogue such as this may only adversely affect the perceptions of that supervisor.

On the other hand, no significant results were found for the Reward, Referent, and Legitimate Position conditions, implying that both humorous and non-humorous statements were rated equally effective in these techniques. One can assume the friendly messages conveyed in the Reward and Referent conditions were already perceived as warmer or more effective because of the promise of a reward or the identification at a common level with an authority figure. As a result, there was no difference between the humorous and nonhumorous conditions because the friendly, comforting effects of the humorous condition might have already been achieved. However, questions still remain about why the Legitimate Position non-humorous condition was not rated as being more effective than the humorous statement for the same reasons as were stated for the Coercive and Expert statements. The belief is that Legitimate Position power is every bit as authoritarian as Coercive and Expert powers and should have been similarly evaluated. Further research should be pursued to investigate the full effects of this condition.

Limitations

Certain limitations of the experiment may have influenced the results. First, because of the design of the experiment, there was no control for the type of humor used in the different power conditions because, obviously, the same humor statement could not be used in every power condition. Thus, some conditions were perceived as funnier than others. More importantly, although the humor conditions were rated as significantly funnier than the non-humorous conditions, the means of the humorous conditions were rather low with a range of 2.14 to 4.00 on a scale where a score of 7 was the funniest score. As a result, it may be possible to assume that the attempts at humor were seen as examples of bad humor. Thus, the experiment may in truth be analyzing the effects of bad humor on bases of power. The findings that the humorous Coercive, Expert, and Legitimate Dependent statements were less effective might indicate the fact that poor humor, not humor in general, is responsible for the less effective rating. As a result, it might be assumed that if one wishes to use humor in attempting to influence a subordinate, it is important to verify beforehand that the humor statement is, in fact, humorous or else the intended results of a more effective leadership style might backfire.

The second limitation of the current study is the fact that participants were not given a definite image or conception of the supervisor upon which to base their perceptions. In the debriefing discussion following the survey, many participants expressed a concern that they pictured an image of their most recent supervisor from their own personal jobs when they evaluated the power techniques because they were not given a concrete description of the supervisor. This was important because many participants were using images of previous supervisors, some of whom may not have been very good at using humor or were not very friendly to subordinates. As a result, the participants' past experiences played a crucial role in rating the individual influence attempts.

As an interesting side note, a number of women participants mentioned that they felt the supervisor was "coming on" to them in the influence statements, even though there was no mention of the supervisor's gender. This perceived harassment may have affected their ratings. Once again, this problem may have been caused by the lack of a visually defined supervisor to control for such effects.

Suggestions for Improvements

Due mainly to concerns and comments stemming from a lack of a clearly defined supervisor, a change in the scenario of the experiment is needed to examine the current topic further. I propose that a visual image of the supervisor be presented to the participants. A videotaped image of one supervisor acting out the individual power statements in either a humorous or non-humorous manner exactly similar to the statements from the surveys for each of the six power conditions could be shown to the participants, who could then rate the effectiveness of the supervisor using the same methods as they used in the current study. In so doing, the supervisor's gender appearance situation, and manner could be controlled. An interesting addition to this new study would be to add conditions of different supervisors' gender and ethnicity to determine whether there would be any significant differences between how different supervisors would be perceived in the separate power conditions.

CONCLUSION

Upon reviewing the research and results from this study, we are struck by one realization: a supervisor's effectiveness is going to be primarily determined by the respect shown by his or her subordinates. In other words, a respected supervisor will be rated as more effective if he/she uses humor in influence attempts because the workers will understand the supervisor and not question his or her effectiveness, even if the humor attempt is unsuccessful or is seen as bad humor. Also, each power base should be more effective if it is used by a respected supervisor because workers would understand the reasons why the supervisor said what he/she did and could change their behavior out of respect for the supervisor. However, it becomes nearly impossible to determine which qualifications would gain the respect of workers in a workplace environment, so our energies should be focused on determining the effectiveness of certain aspects of leadership in the hope that a comprehensive understanding of influence techniques might lead to a better management style.

REFERENCES

American Psychological Association. (1992). Ethical principles of psychologists and code of conduct. *American Psychologist, 47*, 1597–1611.

Bradney, Pamela. (1957). The joking relationship in industry. *Human Relations, 10*, 179–187.

Berkowitz, L. & Daniels, L. R. (1963). Responsibility and dependence. *Journal of Abnormal Psychology, 66*, 429–436.

Coser, Rose. (1959). Some functions of laughter. *Human Relations, 12*, 171–181.

French, J. R. P., Jr. & Raven, B. H. (1959). The bases of social power. In Cartwright (Ed.), *Studies in Social Power*, 159–167. Ann Arbor, Michigan: Institute for Social Research.

Litman-Adizes, T., Fontaine, G., & Raven, B. H. (1978). Consequences of social power and causal attribution for compliance as seen by powerholder and target. *Personality and Social Psychology Bulletin, 4*, 260–264.

Markiewicz, Dorothy. (1974). Effects of Humor on Persuasion. *Sociometry, 37*, 407–422.

O'Quin, K. & Aronoff, J. (1981). Humor as a technique of social influence. *Social Psychology Quarterly, 44*, 349–357.

Raven, B. H. (1983). Interpersonal influence and power. In B. H. Raven and J. Z. Rubin, *Social Psychology*, 399–444. New York: Wiley.

Raven, B. H. (1992). A power/interaction model of interpersonal influence: French and Raven thirty years later. *Journal of Social Behavior and Personality, 7*, 217–244.

Roy, Donald F. (1960). Banana time—job satisfaction and informal interaction. *Human Organization, 18*, 158–168.

Sykes, A. J. M. (1968). Joking relationships in an industrial setting. *American Anthropologist, 68*, 188–193.

Ullian, Joseph A. (1976). Joking at work. *Journal of Communication, Summer*, 129–133.

Wimmer, Ansgar M. (1994). The jolly mediator: some serious thoughts about humor. *Negotiation Journal, July*, 193–199.

ABOUT THE AUTHOR

Michael S. Walters is a recruiter and trainer for the International Business Network, a human resources outsourcing firm in Beverly Hills, California. He has extensive training in the field of social influence strategies and uses these skills while specializing in maximizing business efficiency and employee-managerial relations.

Mr. Walters received his bachelor's degree in psychology from the University of California at Los Angeles (1997). He is a member of the American Society for Training and Development and Professionals in Human Resources Administration, and is active in their organizational development workshops. Mr. Walters has goals of furthering his career in the training and development field.

Mr. Walters can be contacted by mail at 1314 17th Street #29, Santa Monica, CA 90404; by telephone at (310) 315-2736; or by e-mail at mwalters@aol.com.

Author Index

Blake, O., 1
Brantley, C., 10, 19
Bryant, L. E., 34
Buckner, K., 36

Cerri, S., 47, 54
Cranston, R. L., 59
Cutadean, R., 65

Eckstat, A., 74
Eggers, W. C., 81

Gaynor, G., 89
Gess, A. H., 97

Harper, J., 109
Hoschette, J. A., 115

Krause, B., 125

Leech, T., 130
Lillie, J. V., 137
Lockhead, S., 144

McClure, G. F., 151, 158

Paul, R. M., 30
Pierce, D. J., 166

Vavra, L., 171
Voegtli, C., 89, 175, 191

Walters, M., 201

T - #0131 - 101024 - C0 - 234/156/13 [15] - CB - 9780879423247 - Gloss Lamination